MAKING METAL CLOCKWORKS

Stan Bray

Published in the UK by Special Interest Model Books, 2003

This edition published in 2020 by Fox Chapel Publishing, Inc., 903 Square Street, Mount Joy, PA 17552.

Fox Chapel Publishing Project Team
Technical Editor: George Bulliss
Editor: Katie Ocasio
Layout: Christopher Morrison

ISBN 978-1-4971-0059-6

The Cataloging-in-Publication Data is on file with the Library of Congress.

Printed in Singapore
First printing

Contents

Introduction

Clockmaking appears to hold a fascination all of its own, particularly among model engineers, many deciding to make a clock after having made models of various types. There is something absolutely fascinating about clockmaking that seems to draw one toward it. This book is intended as a brief introduction to the tools, materials, and methods generally used and to offer an explanation of general forms of construction. It is not a book of plans, but those who have sufficient confidence could use the information it contains to make a simple clock.

Although clocks come in all sorts of shapes and sizes, the basic principle behind a mechanical clock has not changed for about five hundred years. Of course, modern materials and tools have superseded some of the older ones, but apart from this, the horologist will still tend to work in the traditional fashion.

Clockmaking has long been part of the model engineering hobby rather than being entirely the preserve of the horologist, and the type of workshop owned by the average model engineer is quite suitable for the work. Although a special workshop is not required, one thing that will not do is to work in a workshop that is full of swarf and other rubbish. Accepting that we cannot always have a clinically clean place in which to operate, particularly if it serves several purposes, it is still possible to make a clean area for special use, and this should be a priority. A corner of the workshop can be cleaned and any oil or grease lying on the bench swabbed off. Keep this area clean while clockmaking operations are in progress. It is a good idea to make a false top for the workbench and cover it with baize or a similar material that is soft and will not cause damage to polished metal.

As with all new projects, do not try to run before you can walk. Don't start by trying to build a complicated mechanism such as a full Westminster Chime, but rather make something simple. A mechanism with a single hand is a good idea; such a piece when well polished can look attractive as well as being fascinating to watch when it is working. Visit museums where clock movements can be studied and note how they work. There is actually nothing very complicated about clocks, and most people should soon grasp the basic principles of construction.

All we are looking for when making a clock is a means of making a spindle with a hand attached to it rotate at a given speed. The spindle needs to be driven by some form of power, usually a very primitive form. A means to regulate that power is also required, in order for it to run at a particular speed. If you are hoping to finish your first efforts in this side of the hobby with a masterpiece that neither loses nor gains more than a second or two a year, then you are probably going to be disappointed. It will be possible to achieve a reasonable degree of accuracy and, more importantly, to discover how to improve that accuracy as a result of what has been learned.

When one purchases a very old clock, the chances of it ever being regulated to keep accurate time are generally very small. In this day and age, we are used to being able to buy very cheap clocks almost anywhere that are remarkably accurate. They are controlled by what amounts to a computer chip, and this is how that accuracy is obtained. These timepieces are what one might describe as soulless—they do not have the fascination of the mechanical device—although it must be admitted, they do their job perfectly. The fact that we cannot get this high degree of accuracy does not mean that our clock will be outrageously inaccurate and, after all, for many years it was quite customary to set a clock to the correct time once every week or so.

Seen at an exhibition at the British Horological Institute were these clocks made basically from Meccano.

Fortunately for those who are beginning, the material that will be used is neither extensive nor expensive when compared to the castings, etc., used for model engineering purposes, so do not be frightened to make a part more than once if something has gone wrong. It is far better than trying to recover something that has not been correctly made in the first place. Unlike building model locomotives or traction engines, where there are expensive castings to be bought, some brass sheet, a few bits and pieces of drill rod, and we are in business.

Although we generally think of brass as the material used for clockmaking, there is no reason why other materials should not be used. The photograph shows a wheel for a clock made from plexiglass (Perspex).

We all tend to think of clocks being made from brass and steel, but other materials can be used. There are a number of plans available for the construction of wooden clocks and, while this may not sound like a suitable material, it is surprisingly robust: the Germans have used it for years to make clocks commercially. Nowadays plastic can be a useful material. It is easy to work with, it is durable, and a clock made of a transparent plastic can be a fascinating thing to see. Also available are plans for clocks made from paper and cardstock (card). They appear to work very well and last for a long time.

Tools

Most of the tools likely to be needed will be found in the workshop of the average model engineer. Needle files, hacksaw, lathe, some small drills, and taps are the basic items that are wanted. In addition, a small five-sided broach is possibly the only essential item that might not already be part of

One tool that will be essential is the five-sided broach as seen here. They are available in a variety of sizes and can be obtained in complete sets. The type shown is for cutting pivot holes; another type is available for smoothing them.

the workshop equipment. There are additional tools, of course, but they are not absolutely essential. Wheel and pinion cutters could be very useful, but it is quite possible to go without them. There are many people making very fine clocks who have never bought such a cutter in their life. A depthing tool (see Chapter 6 for further information) is useful for laying out the wheels, but once more, far from essential; there are several ways of doing the job without such an item and, like cutters, they are easily improvised. Another useful item is known as a Jacot tool; it is used to obtain a high polish on pivots and as an aid to making them. It is quite possible to work without one and, if it should be thought to be an absolutely vital piece of equipment, it is very easily made.

The Lathe

Watchmakers' lathes are expensive to buy and are of little use for purposes other than watchmaking; in fact, they are of doubtful value when it comes to making clocks. Generally speaking, the type of lathe found in the average model engineer's workshop is quite suitable. The most popular of all these are probably the mini-lathes, which have a swing of 7" in diameter. Thousands of good clocks have been made using them and other lathes of a similar size and specification. The miniature-type lathes with center heights of about half that have the advantage of being cheap and, because they are small, obviously less space is required. Most are available with a bolt-on mill/drill attachment that makes them ideal for cutting the teeth on wheels. It is not possible to give advice on the best lathe for a newcomer to purchase—it all depends on a particular

Generally speaking, the tools needed for clockmaking will differ little from those used by a model engineer, although one or two extra items may be required. The photograph shows a piercing saw, which is useful for many purposes, but in particular for crossing-out wheels.

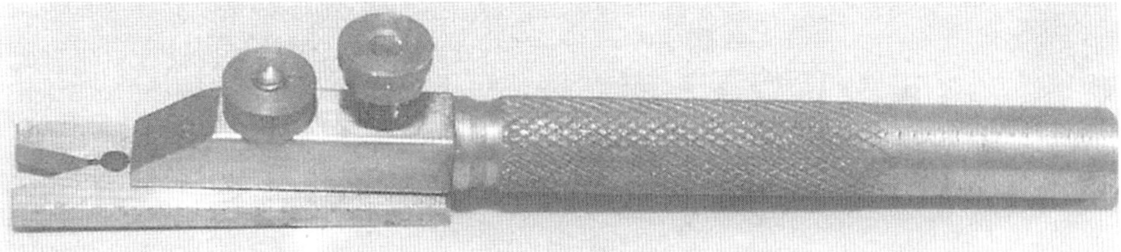

A small hand clamp like this one can easily be made very quickly and is invaluable for holding small parts while working on them.

preference and one's budget. In addition, many of the foreign lathes that are sold are only available for a limited period of time before the specification is changed. Anyone wanting to purchase a lathe would do well to visit one of the model-engineering exhibitions held throughout the country, where it will be possible to browse through a large number of machines of different makes and types to find which might be the most suitable.

Some of the methods used by clockmakers are likely to make an experienced engineer wince. For example, while the engineer will always try and get as much bearing surface for a spindle as possible, the clockmaker seems to try to do the exact opposite. Holes that are to be used for bearings are drilled smaller than the diameter of the spindle that will fit in them, and then they are reamed with a taper broach until a fit is made. As a result, the spindle is running on the thinnest possible ring of brass. To the engineer the idea sounds frightening, but really it makes complete sense. Although the movement or mechanism is rotating continually, it is hardly going to break any world speed records in so doing. It has no real work to do inasmuch as it is only driving itself, and so the small bearing surface has the advantage that it cuts friction down to a minimum. As a result, it reduces the power needed to run things.

Terminology

This is another thing which engineers may find a little difficult to understand. A shaft or spindle is known as an arbor and the bearing surface at the end is not a shaft or axle but a pivot. Gears become wheels in spite of the very obvious teeth all round them, the making of which is the main part of clockmaking. Although the teeth on the wheels are called teeth, when they are on a pinion they are frequently described as leaves. These gears (wheels) are made of thin section brass, which is ideal for the mechanism that is being constructed. In order to give a good fit on the spindles (arbors), they are fitted on bosses, from now on known as collets. To an engineer, a collet is

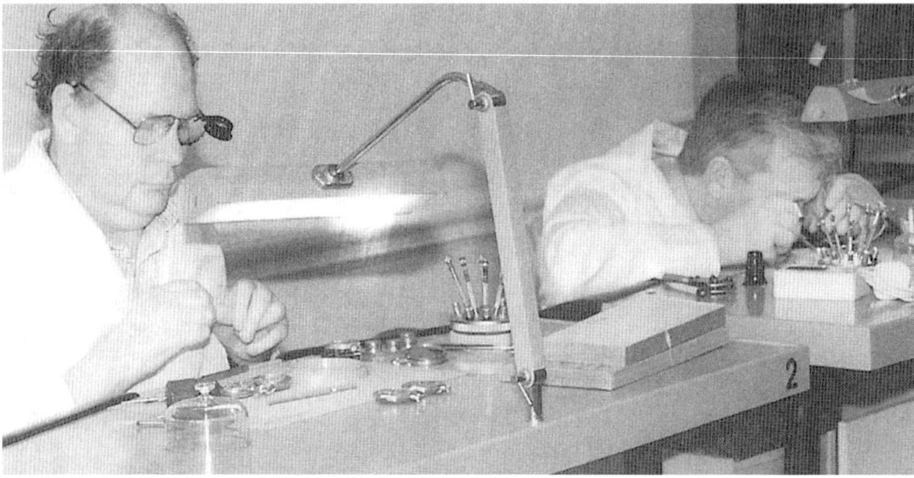

Two members of the British Horological Society at work in the workshops. The photograph gives some idea of the type of bench to aim for. A clean surface, a tray for parts, and a good light are the basics, with some small storage stands for the few tools required.

Generally speaking, any lathe will do for clockmaking. The photograph shows the Cowell, which has been specially designed for the purpose.

something that opens and closes to hold tools or material. To a clockmaker, the term includes the length of brass used to support a wheel. And, finally, there is the mechanism itself, which is called a movement. It is all very confusing at first, but we must remember that every trade has its own terminology—just look, for example, at those used by the computer engineer.

Help and Assistance

There are a considerable number of plans available for making clocks of various types. Some, such as the designs of John Wilding, are sold in book form, complete with full instructions; they are very highly recommended. Other similar books are available as basic plans, and in some cases complete kits can be purchased. The photograph on the front cover of this book is one of a movement made from just such a kit, by Repton Clocks, and this, too, is an ideal way to learn the basics of clock construction. There are also a number of suppliers of clock parts and books based throughout the country.

The British Horological Society stocks a wide range of books and plans. Here we see a photograph of the library, possibly the most extensive in the world on the subject of watches and clocks.

If we accept all the oddities that are imposed upon us, clocks are fascinating. There are no heavy castings to hump around, the work is nice and clean, and the end result is worthwhile. Even if after reading this book the reader should decide that clockmaking is not for him or her, it is still worthwhile taking an interest in them, their history, and the beauty of the finish on many. This in itself can provide an everlasting interest.

This book deals with simple clockmaking techniques; it is possible to progress further to such things as striking clocks or, perhaps, as seen here, an orrery, which gives the rotation and phases of the moon and planets in relation to the world.

Chapter 1—History

Before starting on constructional details of clockmaking, a word or two on the history of the subject may assist readers in an understanding of timekeeping in general. It is not only the human race that uses timekeeping methods; animals know when it is time to go to sleep or, if nocturnal, when to start searching for their food. Some plants and trees will close down flowers at night and open them when daylight appears. Of course, this has nothing to do with clocks as we have come to know them; they are reacting to light and dark and possibly also to changing seasons. No doubt the human race also started in this fashion, sleeping during darkness and active during the daylight. The clocks in use then were the sun and moon, not necessarily the most reliable sources, as, for various reasons, they are not always visible. Such primitive methods, while not exactly telling the time of day, did give reasonably accurate measurement of the seasons, and so had their uses.

It was not all left entirely to chance. Archaeology and ancient manuscripts tell us that the movement of the heavens has been observed for thousands of years and that this movement has been applied to timekeeping methods. The new moon appears every thirty days and the seasons repeat themselves every twelfth time it appears, which forms the basis of the year as we now record it. The equinoxes were well known and used for religious purposes, and stars and sun were used for early navigation, whether across sea or land.

During and prior to the Stone Age (from about 3.3 million years ago to 8000 BC), it is very doubtful if anything more accurate than this would be required. If the sun was at its highest point, then it was halfway through the day. Humans are never satisfied with basics and we can only speculate when it became desirable to be able to split time into smaller parts and exactly how it was done. It seems highly probable that a primitive sundial would be the first basic form of clock. Put a stick vertically in the ground, and when the shadow cast by that stick is at its shortest, it is midday. Who knows, perhaps it was a tree that first gave someone the idea that the shadow of the sun could be used to give an approximation of time.

Sundials did not remain as sticks in the ground, and we know stone columns were used in early times, metal pillars of varying shapes following them. This is not, however, a book about sundials but about clocks. The sundial subject is so vast that it would take a separate book to discuss it. What we do know is that sundials were in use around 200 BC, and a hundred years or so later a geared mechanical device was produced for navigational purposes at sea, which may or may not have been a primitive form of clock. In 600 BC, the Pope decreed that all religious institutions should have a sundial as a means of regulating the times for prayer, so the human race was really becoming more time conscious.

Not all the world's population was Christian, and we must look at those countries that had not adopted the religion to see, where possible, how they sorted things out. It is known that the Chinese had the idea of using water. Although there was a number of variations on the theme, the basic principle was to first fill a container with water. A small hole in the bottom would allow it to run away, and by measuring the amount that had gone, it was possible to see how much time had passed and, if the container was marked with graduations, the time elapsed could be seen at a glance. This type of clock eventually became used all over the world, and various improvements made to the system, including attaching a dial, ensured that the water clock remained in use for hundreds of years.

Candles were also used as a measure of time. Once it could be established how much a candle burnt down in a particular period of time, it was a simple matter to mark the sides, showing how much time had elapsed since the candle was lit. King Alfred the Great (ca. 847–899) is credited with being the first person to use candles for timekeeping. It is still possible to buy candles marked in this way; nowadays they only have novelty value. An almost identical idea to the candle was to burn oil in a container with markings to record the time that had passed as the quantity of oil was reduced. For short time periods, there was also the sandglass, with which many of us will be familiar as an egg timer. The glass and quantity of sand had to be carefully matched to the time required, so the system was generally used only for specific purposes, because indicating intermediate time between filling and emptying the glass was not possible.

What of the religious orders that went to prayer night and day? Something was needed to tell them when it was time to go to the chapel. Even though the Pope had decreed that all religious institutes should have a sundial, this was of no use after dark or in much of the weather we have in Britain.

Various ingenious measuring devices that sounded alarms were devised; these included weights on a piece of string that was set fire to. After a period of time the flame burnt through the string and the weight would drop on a gong, telling the monks that it was prayer time. The idea was extended to include a number of weights strung to a frame. The strings were of different length and so burnt through at different times. In this way the gong could be sounded automatically at set intervals.

Nobody knows when the first mechanical clock came into being, or by whom or how it was invented. The oldest clock known of in Europe was at Salisbury Cathedral and is dated 1386. It is still in working order, and although no longer in the tower, it can be seen in the nave of the cathedral. It is quite an advanced clock that includes a mechanism for striking the hours, as well as one for telling the time. Even this is not the first known mechanical device: in 1090 Su Sung made a device in China that rang bells at given intervals in addition to driving automata, although it did not have a dial for timekeeping. Driving automata was popular with clockmakers, and one early example at Wells Cathedral, built in 1389, can still be seen working. It is a very elaborate affair, with knights jousting and all sorts of other movements at set times. We do know therefore that clocks have been in use for many hundreds of years, and although the construction of early ones appears to be quite crude, many were in fact sophisticated pieces of machinery. Early clocks were used for public purposes and were very large. Generally they would be housed in the tower of a church or cathedral.

There are records of clocks for domestic purposes as early as 1343 in France and England. The earliest surviving examples date from around the fifteenth century and are of iron construction, the mechanism being scaled down from the larger ones in public places. All these early clocks were weight driven, and exactly what date the pendulum replaced the folio control we cannot say. However, a claim is made that a clock was made with pendulum control in 1656. Spring drive is first heard of in 1450 and therefore predates the use of the pendulum. Regular improvements were made to the timekeeping mechanisms, including, in particular, the invention of new more reliable escapements, and in 1715 George Graham invented the deadbeat escapement, making clocks more reliable still.

It is quite amazing to think that modern mechanical clocks work on exactly the same principle as they did when George Graham invented his escapement. Materials have improved, with brass and steel taking over from iron, but otherwise there is little difference in the basic construction of any type of mechanical clock. Mass production was really the only big advancement from then on, but minor improvements to both clocks and watches continued.

In the twentieth century, clocks and watches saw advancements that would not have been thought possible, even at the start of the century. About sixty years or so ago, people first started to experiment with the use of electric clocks. Certainly, they were very primitive in comparison with what was to come later, but they had the advantage in that it was possible to synchronize several clocks together, which was ideal in a factory where many people were employed and all would be starting and stopping work at the same time. Prior to that, a bell or hooter was used, a system that in many places lasted until quite recent times.

Improvements in the manufacture of electric clocks were rapid until we reach the stage at which we are at today, where it is possible to buy a clock with a digital readout so that nobody even needs to know how to tell the time anymore. The modern electronic masterpieces can keep perfect time and are far more accurate than any except the most expensive mechanical clocks. Probably because of their efficiency they do not have the fascination of mechanical ones, which are still made today both commercially and by amateurs.

Many people buy mechanisms for these quartz clocks and make cases of various types to house them; it is a hobby on its own, in which large numbers of people are happy to indulge.

Before finishing with the history of clocks, it is interesting to think how time itself has changed. Until quite late in the nineteenth century, every town or district kept its own time. Communication between areas was very poor, with limited transportation, and it mattered not what time it was in a town forty or fifty miles away. With the coming of the railways all this changed. A person traveling from say London to Birmingham and then wanting to get a connection to somewhere else needed to know what time that connection would leave in relation to the train on which he or she would arrive. The railways therefore organized their own time, known as Railway Time, which was consistent right throughout the country. Gradually this was adopted throughout the country until everyone used the same. Now time is related directly to the Greenwich Meridian and known as Greenwich Mean Time. Other countries also take their time from the meridian, with allowances made for time zones. As a result, it is possible for anyone, anywhere, to know what time it is in any other part of the world.

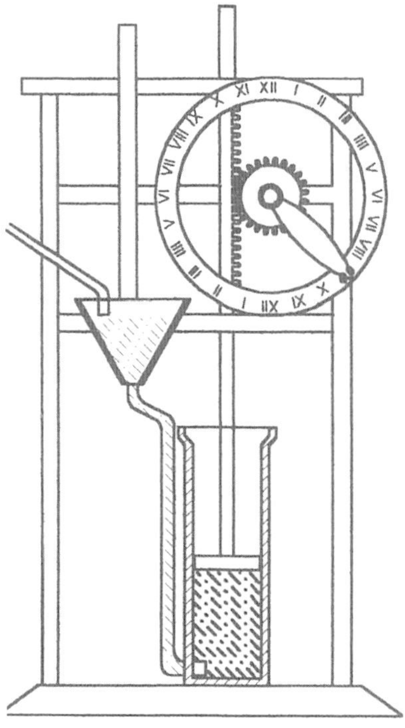

The drawing represents an ancient water clock of about 200 BC, said to have been made by Ctesibius of Alexandria, who was a famous inventor. Water passes through the funnel into the reservoir, and in so doing raises the plunger. This incorporates a rack, and a hand with a gear wheel attached rotates to indicate the hours. The carefully regulated water supply is allowed to run for twenty-four hours and then the reservoir is emptied and the cycle repeated.

Chapter 2—The Frame

The frame of a clock will generally be made of two flat plates joined together at or near the corners with pillars. All parts are usually made of brass except in exceptional cases where we might get a steel frame fitted with bushes. The plates are sawn and filed to size, and, after ensuring they are flat and square, they should be held firmly together with clamps, preferably the toolmakers' type, while two or three small holes are drilled through somewhere near the corners. These holes are to accept pins or rivets that are used to ensure the plates do not separate during operations; once the pins and rivets are in place, the clamps can be removed.

The next task is to mark the position of the pillars that join the plates together and drill the holes for them; we will come to how they can be fitted shortly. Occasionally clock designs do not have this type of plate; instead they are made with strips of brass, more often than not cut into fancy shapes, and instead of four pillars there are only two, one at each end. The principle of joining them together and drilling the pillar holes remains exactly the same, however. As building progresses, other differences will emerge. For example, there will not be a pendulum, so they will not be fitted with a back cock.

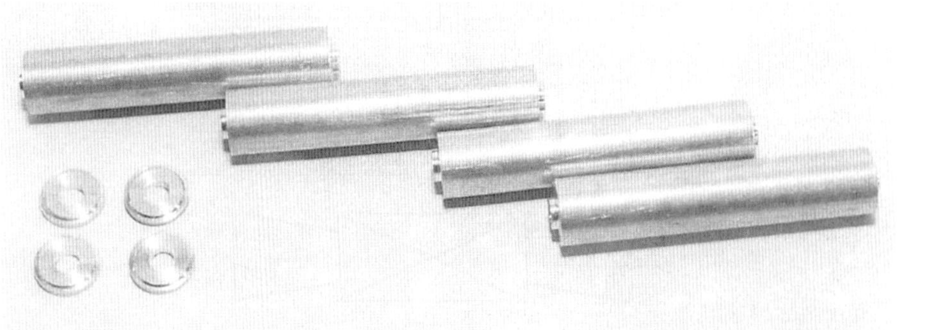

Four basic pillars. There are several ways of securing them and, in this instance, locating lugs have been machined on the ends, which are tapped to accept screws.

The Pillars

Generally speaking, the pillars, or spacers as the layman would call them, will consist of brass bars that may or may not be shaped. Shaping is a matter for the individual builder and, in a limited way, is the opportunity for him or her to express him or herself. Fitting the pillars to the frame is done in several ways: some are hollow and a stud is pushed right through and the parts held secure with a nut, or perhaps the ends of the pillars are machined down and threaded to accept a nut. In other cases they are drilled and tapped and screws passed through the frames into them. A third alternative is to machine a step in the pillar ends and pass this through the holes in the frames, which are then secured with a taper pin, fitted in a hole drilled across the step. One thing that is common to all these methods is that, when assembled, the frames must be rigid and square.

The two plates held together with the four pillars and with the barrel fitted for test purposes.

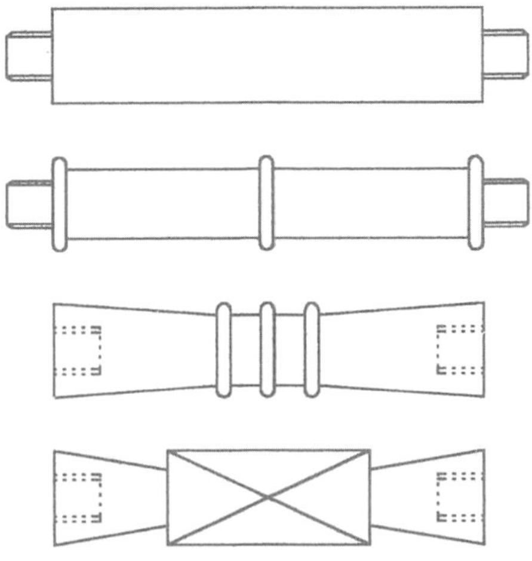

Pillars should generally be made of brass, unless another material is considered more suitable. The securing threads can be external for use with nuts, or internal for screws. The shape is only limited by the constructor's imagination; some suggestions are shown above.

Setting Out the Train

The most common way of setting out the train is to scribe a straight line lengthwise down the plates and to set the escapement, center or hour wheel, and the great wheel and barrel along this. The third wheel has to be set at one side in order to allow the pinions and wheels to mesh. Just occasionally we come across another design where the escapement and hour wheel are in line and both the third wheel and barrel offset. This is very rare and any details required for such an arrangement would be available from the drawing and any instructions that might go along with it.

Marking Out

Sometimes clock plans will give measurements showing where pivot holes will be placed; if not, it will be necessary to work out spacing for oneself. Start by lightly dot punching a suitable place for the great wheel on the centerline. Use a depthing tool to mark out a position on the line of the minute wheel; this means meshing the great wheel pinion with the minute wheel so they run very smoothly and without any binding. When satisfied with the meshing, use the tool to make a second mark on the line that has been marked on the plate.

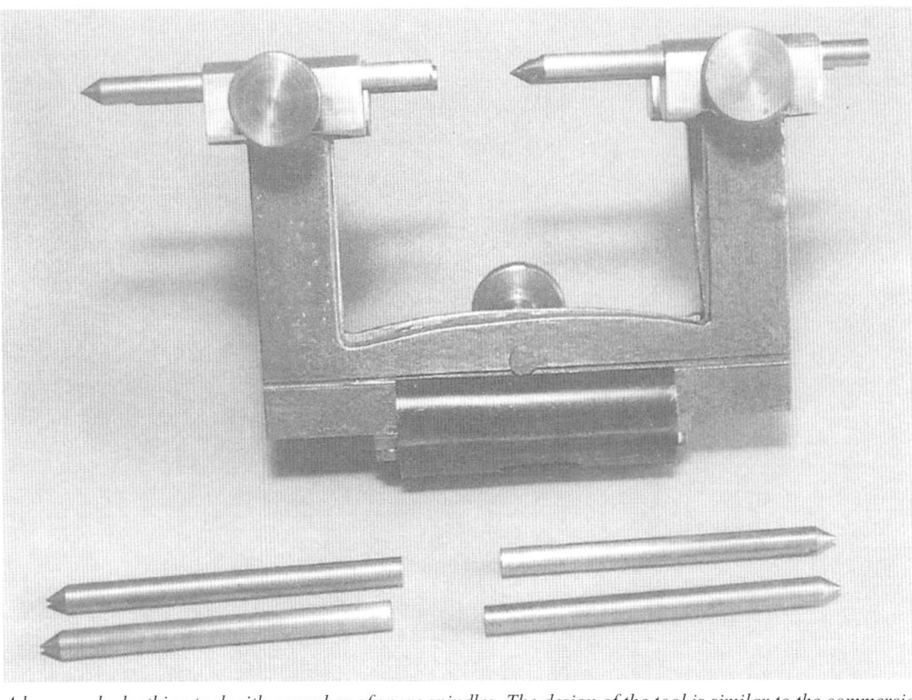

A homemade depthing tool with a number of spare spindles. The design of the tool is similar to the commercial models and requires the use of a heavy spring.

Depthing Tool

A depthing tool is something that some people will not have come across before; it is a tool for setting out gears to ensure that they run smoothly. They can be bought, but for normal purposes a homemade device will do just as well, those that are purchased being far more sophisticated than necessary for occasional clockmaking purposes. The tool is simply a means of meshing wheels and pinions, or two wheels, or the escape wheel and pallets, so that a check can be made to ensure they run properly. A professionally made tool will be spring-loaded and fully adjustable, but good results can be obtained from a simple device consisting of two lengths of bar that swivel together with two holes to accept punches. The punches are the same diameter as the wheel arbors, and so the wheel and pinion are simply slipped on and adjusted. If different-sized arbors are likely to be used, fit brass bushes that can be interchangeable. One of the punches is set in the mark already made and the other is lined up on the line on the plate. A slight tap with a small hammer and the correct place for the arbor of the hour wheel is marked. John Wilding, who is one of the finest clockmakers in Britain, recommends a piece of slotted bar for the same purpose, an idea that works very well.

The back cock must be absolutely square, otherwise the pendulum cannot operate properly.

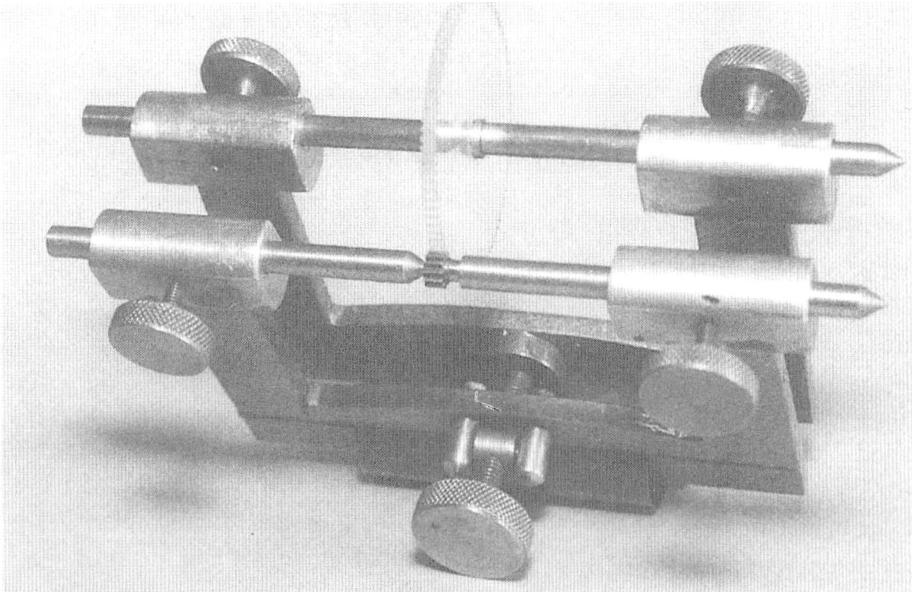

The depthing tool being used to assemble a plexiglass (Perspex) wheel and a pinion. The tool is adjusted until they are running freely and then it is locked in position. The tool is placed on end, and a smart tap with a hammer on each spindle leaves indents at the correct spacing ready for drilling. As well as obtaining the distances for wheels, the tool can also be used to check the working of the escapement.

A much simpler depthing tool consisting of two pieces of strip steel fitted together with a screw. The spindles are the right size for the wheels, and when they are slipped in place, the necessary adjustments are made. The spindles act as punches to locate the hole positions in the plates.

The tool is now used to set the distance of the third wheel, which cannot be sited on the line because it would then be impossible to mesh all the wheels properly. There is no specific angle at which to set the third wheel and this, plus whether it should be set to the right or left of the line, is a matter of individual taste. Generally speaking, an angle of about forty-five degrees is used; whatever happens, make a note of the angle just in case it needs to be referred to later. Having decided where the wheel will be going, repeat the operation with the depthing tool so that a mark is made for the third wheel pivot hole. From there the tool is used to lay out the distance from the third wheel to the escape wheel, and exactly the same procedure is used. The mark should be made on the line used by the great wheel and minute wheel, and that is the clock train marked out. Holes for the pivots can be drilled, but don't forget they should be drilled undersize. When the plates are separated, they can be opened out with a taper reamer so they are a nice running fit with the pivots.

There is still one more operation required in order to make all the pivot holes, because the pallets of the escapement also need to be set. The pallets and the escape wheel are mounted in the depthing tool and adjusted until they work smoothly in the same way as before.

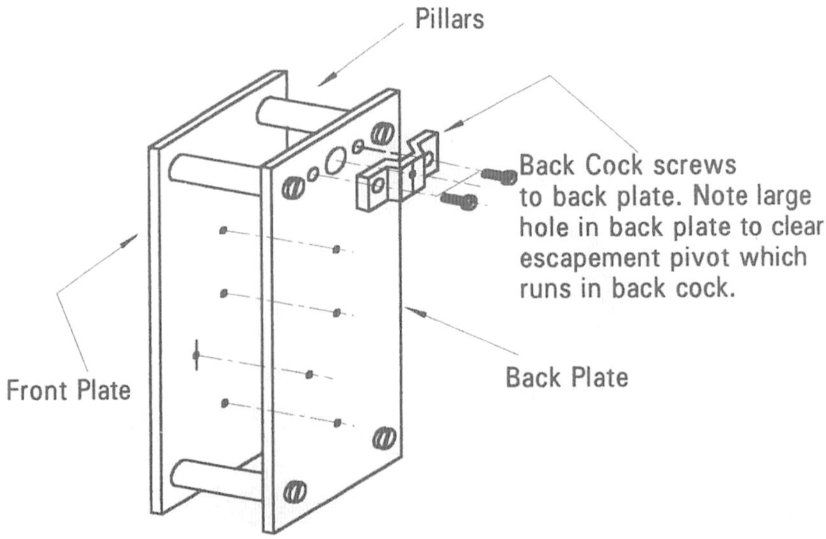

Makeup of the plates, showing typical position of pillars and back cock.

The plates are now complete, except that if the clock is to use a pendulum, a back cock must be fitted. This is a bracket that supports the suspension for the pendulum, which can be fabricated or milled from a solid block of brass. It takes the form of a bridge to clear the pallet arbor, the pivot of which locates in a hole in the cock, the suspension being fitted on the arbor between the plate and the bearing. It is essential for the good running of the clock that the cock is secured firmly to the plate and will not work loose. It is also essential that the escape wheel arbor when fitted to the cock is at ninety degrees to the plates.

Some people like to fret out the plates to make an attractive pattern, and this is a matter of individual taste. The fretting will obviously need to be done with the plates secured together. One way of doing this is to draw a suitable pattern on paper, taking care to avoid all pivot holes, cut out the pattern, stick it on the plates, and cut around it. Do not try and chain drill, but drill a couple of holes in strategic places, and use a piercing saw or a power-operated scroll saw, if one is available, to complete the job. Finally, clean the edges of the cuts using small files.

The back cock in position and the crutch temporarily in place.

Something rather unusual in back cock construction, consisting of four spacers and a plate, which is bushed to accept the escape wheel arbor as the plates, in this instance, are made of steel.

The plates in position and the movement assembled.

Chapter 3—Providing the Power

Most people, when starting clockmaking, begin with a weight as the power source, and so we will look at how this is done first of all. A suitable barrel is required, around which a line can be wound. This in turn is tied to a weight, but not directly because the weight is connected to a pulley and the driveline passes through the pulley and is anchored on or near the clock. The usual place is one of the pillars. Any tubing will do for the barrel, although it must be thick-walled. Generally speaking, brass is preferred, mainly because of its non-rusting properties, but there is no reason why it should not be of steel; many years ago most clocks were made of iron, although the use of that material is not to be encouraged unless trying to build a replica. A spindle, or arbor as horologists prefer to call it, is passed through the middle and on one end of the arbor is a square. This is to accept a key with which to rewind the line when it reaches a low point.

The barrel is made from thick tubing, the ends of which must be machined square.

The Barrel

One end of the barrel is plain, except for a hole through which the gut is secured. The other end forms a ratchet, referred to as the click wheel; this can either be made integral with the barrel or fitted permanently in place to an end cap. How the end caps are fitted is a matter of personal choice, but there is much to be said for using small screws to hold them in position, as it enables it to be dismantled should any work be needed on the arbor. Also connected to the arbor is a wheel, or gear if you prefer, called the great wheel. This connects via a pinion to the train, and the ratchet prevents the barrel from unwinding, except when controlled by the rest of the clock. To locate in the ratchet is a specially shaped metal plate, known as a click, and a spring to hold it in position. Although the barrel can be left plain, it is a good idea to make a continuous groove in it to allow the gut to wind on and off evenly, rather than pile up. A small round-ended tool, similar to a parting-off tool, is ideal, and ordinary screw-cutting methods can be adopted. The depth of the groove should be about half the diameter of the gut being used, and about seventeen or eighteen turns are required. Therefore, if the barrel is 2" long, the lathe needs to be set to cut nine threads per 1", or the nearest available figure above that.

The Ratchet

The ratchet can again be of steel or brass, although brass is generally recommended, and the teeth are cut in the same way as on all the wheels. The number of teeth varies considerably, depending on the design of the clock. In some instances, designers have deliberately kept the number of teeth on the ratchet to a minimum. This allows it to be filed to shape, avoiding the necessity to make or buy a specially

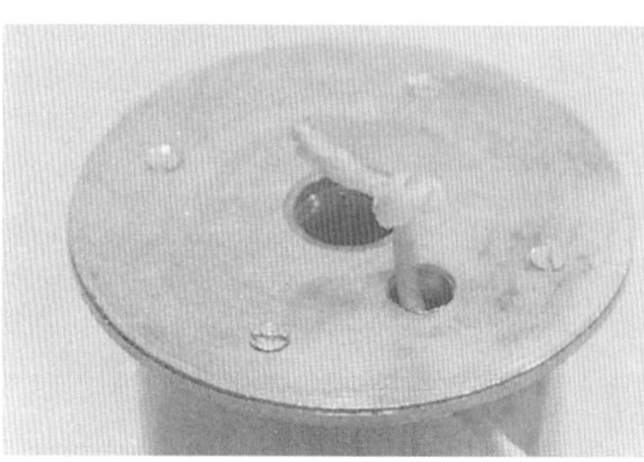

The cord for winding is traditionally made from catgut and passes through holes in the barrel at the opposite end to the ratchet, where it is tied in a knot to prevent it from slipping through.

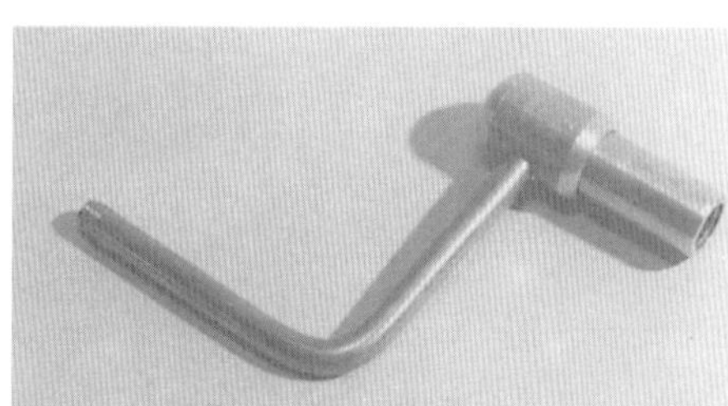

A key will be needed for winding up; it can either be handmade, as in this case, or purchased. Many that are bought have attractive shapes on the handles. The only difficult part when making a key is the internal square section, which has to be filed to shape.

shaped cutter. In these instances, the teeth are curved in a gentle radius, both to facilitate filing and to make the ratchet smoother in use. The more normal type of ratchet will have about thirty-six or so teeth, and cutting by machine is almost essential. This is not to say that the work cannot be done by hand, but a high degree of skill is required to get all the teeth to the same profile, not to mention that the task will take some considerable time.

The Great Wheel

The great wheel is mounted on the barrel arbor and drives the train via a pinion and, apart from cutting the teeth, it is quite straightforward. As the weight unwinds, it drives the great wheel, and when the clock is being wound up, the ratchet allows it to slip past.

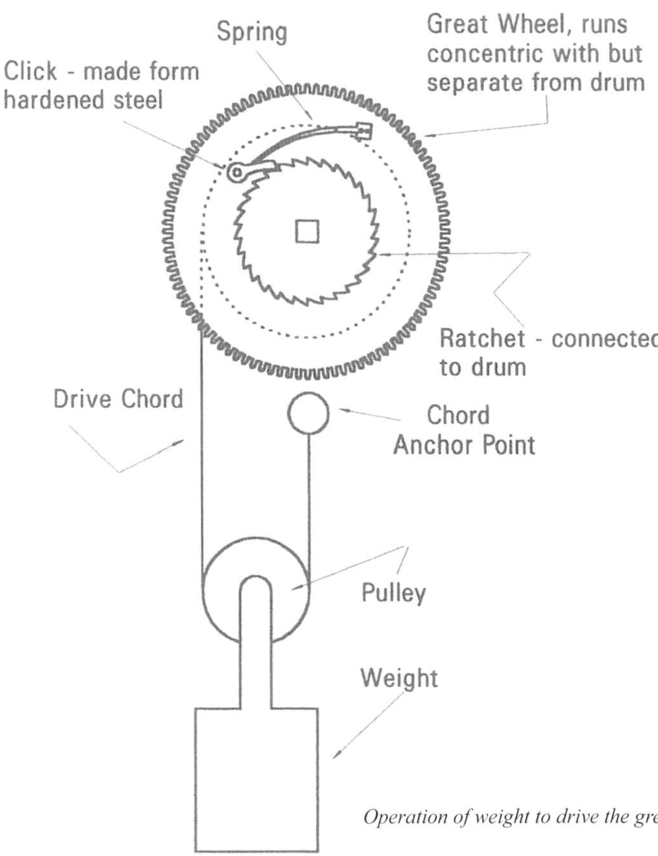

Operation of weight to drive the great wheel.

The Click

This is the name given to the pawl that locates on the ratchet and prevents it from unwinding. It is usually filed to shape and the shape is not difficult to make, the only critical part being the distance between the hole center and the tip of the blade. It should be made from gauge plate and hardened, then tempered to a dark blue color. As an alternative, it could be made from mild steel and case hardened and, while not giving quite as good a result as the previous method, a long-lasting and reliable click would still be the result.

Click Spring

The spring holds the click down on the ratchet wheel and must therefore be strong enough to do so, while at the same time not being so powerful as to drag and cause more power to be needed than would otherwise have been so. The springs vary considerably in length and shape and so materials may vary with different springs. Mild steel can be used for most springs, and although it may sound a most unlikely material, hammering thin sections causes a work-hardening effect, resulting in springiness. Another useful material that

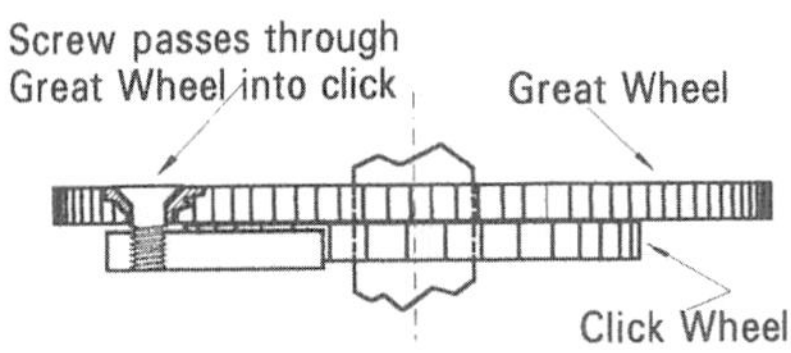

Drawing showing how the click is screwed to the great wheel and mates with the ratchet wheel.

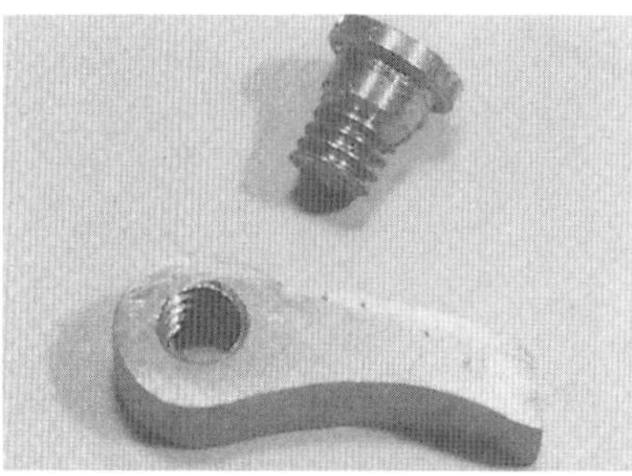

The ratchet click and securing screw. The click has been made from gauge plate and hardened; case-hardened mild steel would do just as well.

Suggested shape for click, which locates with ratchet on the drum.

Shape of a typical click spring.

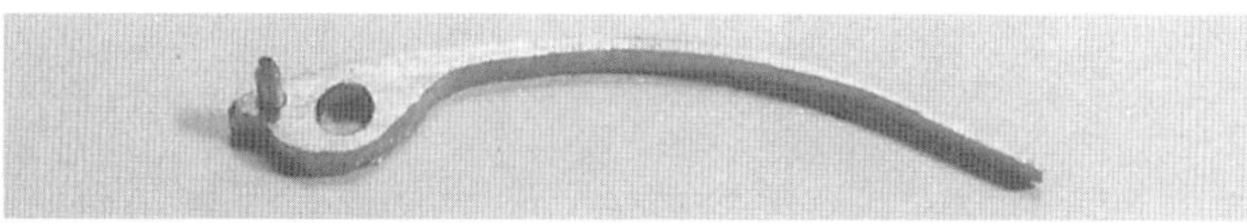

A click spring made from mild steel and hammered until the work-hardening effect gives it a suitable spring action. The pin locates in a hole in the ratchet and, together with a screw that passes through the hole, acts as a securing device.

A ratchet and click assembled, showing the action of the spring.

An arbor for a barrel. It is square at one end to accept the key with which to wind the clock. The other end has a groove, which takes a spring washer, allowing the arbor to rotate independently of the rest of the train when the clock is wound.

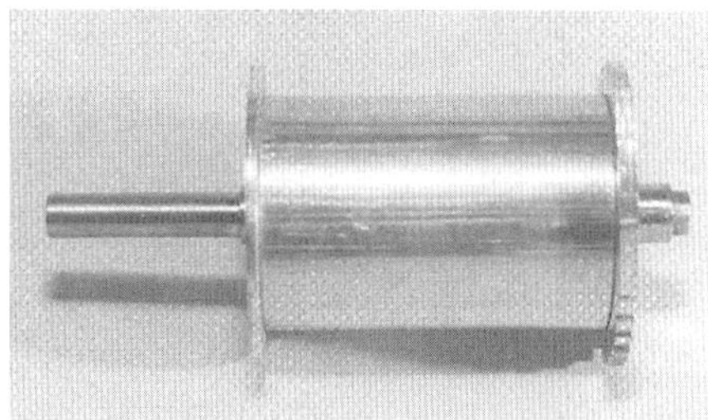

An assembled drum with ratchet.

Two caps are made for the ends, one of which has the ratchet wheel fitted to it. In this case it is screwed in position, but it could just as easily be made as a single piece. The end caps can also be screwed to the barrel.

has a similar effect is drawn phosphor bronze, a material with a natural spring to it and which is also very easy to work. Like all parts for clocks, suitable springs can be purchased completed and ready for use, if one wishes.

The Pulley

The weight that will drive the clock is suspended from the drum via a pulley, which effectively halves the weight applied. There is not a great deal that can be said about pulley construction; we are all familiar with the shape of the wheel, which should generally be of brass and run on a steel axle. The frame of the pulley can be made from a piece of brass plate, and that is all there really is to it. The appearance can be improved by drilling holes in the pulley wheel.

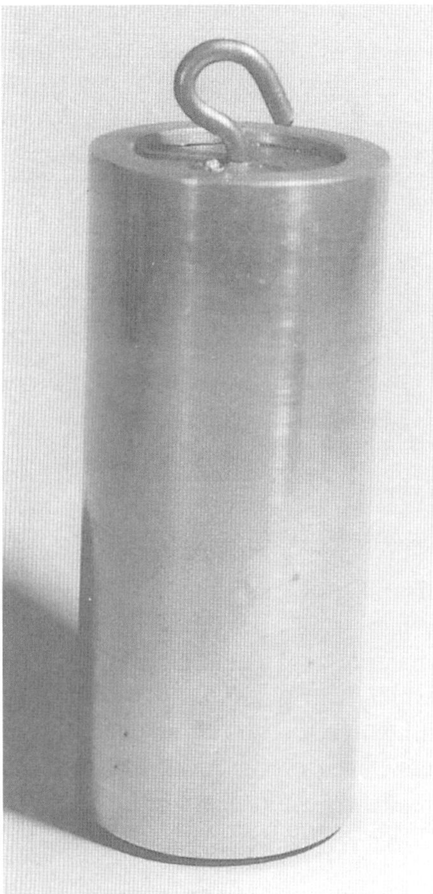

A weight for driving the clock can be made by pouring lead into a brass tube, as in this instance. The top cap with the hook has been added after filling the tube and soft soldered in position.

The Weight

Various materials are used as weights, which are also made in a variety of shapes. A nicely polished length of brass tube filled with lead is ideal; do remember that a hook is required with which to hang the weight on the cord and that this must be set centrally to the diameter so that the weight hangs straight and does not lean at an angle. Although the weight is mentioned in this chapter, because it is part of the driving mechanism, it will be one of the last items required. It is obvious that we cannot hang any old weight on the clock and expect it to keep time. It must be sufficient to keep the clock going without causing it to work at too fast a rate, so the amount of weight required will be a matter for experiment.

On smaller clocks, a weight drive is not practical and a spring is used, preferably in conjunction with a fusee. The spring will arrive tightly bound with wire as shown here. Fitting it into the barrel is difficult, but not impossible. Even so, it would be advisable to get the local clock restorer to do it.

A good idea is to get a used food can and fit to it some means of attaching it to the line from the drum, fill the can with pieces of lead or other heavy material, and run the clock. Keep removing small amounts until the clock stops. This has to be done over a period of several days. When it has stopped, weigh the contents of the can, add about half again, and make the finished weight to this measurement.

Spring Drive

So far we have dealt only with clocks that are weight driven, which means they are either fitted in a long case or hung on a wall. What if we want our clock to sit on a shelf? It is hardly practical to drill a hole in the shelf and run a cord through that to a weight. The answer is to drive the clock with a spring, something with which we are all familiar and which is freely available. They arrive coiled and sealed tightly with a fastening almost ready for use, and fit inside the barrel using two hooks, one of which attaches to a point inside the barrel, the other to the arbor. Great care must be taken when dealing with springs, as they can cause nasty injuries, and it is advisable to wear heavy gardening-type gloves and most definitely to wear protection for the eyes. Special devices are available for setting springs in barrels and, while it is possible to do so by hand, if in doubt, find your nearest clockmaker and take the lot along there, where a special device will be available which will enable the job to be done in a matter of minutes.

Generally, the barrel will differ considerably from that described for the weight-driven clock. For a start, it becomes obvious that one end must be removable in order for the spring to be inserted, but other major differences also occur. Although not unknown for a clock to be driven directly by a spring, particularly if one buys a cheap one, it is most certainly not good practice. While the weight drops at a given speed throughout its length, the spring behaves very differently. When tightly coiled it creates considerably more power that when it is only partly wound. Watch a clockwork-driven toy and see how it slows down when the spring starts to run down, and of course that is something that is not wanted in a clock.

The Fusee

To avoid this problem, it is usual to connect the spring via a device known as a fusee, which is a tapered and grooved length of brass on an arbor, on which is set the great wheel. It is much the same as the situation with the barrel on a weight-driven clock. A cord is wrapped around the grooves in the barrel and runs to the fusee. When the spring is fully wound, the chord passes around the smallest part of the fusee, effectively acting as a brake. As the spring loses its power, so the chord winds to a larger diameter, keeping the rotation of the fusee at an even speed. The arbor on which the fusee runs is similar to that of the barrel on the weight-driven clock, and the barrel now works independently of the clock movement, providing the power and nothing else. The clock is wound via a square on the fusee arbor, which in turn rotates the barrel. As the spring is hooked inside that, it tightens up, to be released at a regular speed, with the fusee acting as a continuous gear and compensating for the unequal torque of the spring. To enable the fusee to be wound, the conical-shaped part is not directly attached to the great wheel but drives it through a ratchet as already described for the barrel. From there the power is transmitted to the rest of the train.

Fusee Construction

Many people avoid making the fusee and prefer to purchase it, but it is a task that is well within the ability of most people. Take a length of brass bar of a slightly larger diameter than the maximum diameter of the fusee, put it in the three-jaw chuck, face the end, and drill a hole through its length for the arbor. Machine the outside diameter to size, and we can now guarantee that the hole for the arbor is true to the outside diameter. Set the top slide over and machine the required angle, then use either a radiusing tool or a hand graver to generate the required curve, which must be smooth.

So far it has all been easy going, but now we come to the only tricky part, which is to machine the continuous groove. A setup for coarse screw cutting will take care of the spacing, which is the same as that for the barrel, so all we need is to organize a tool that can be

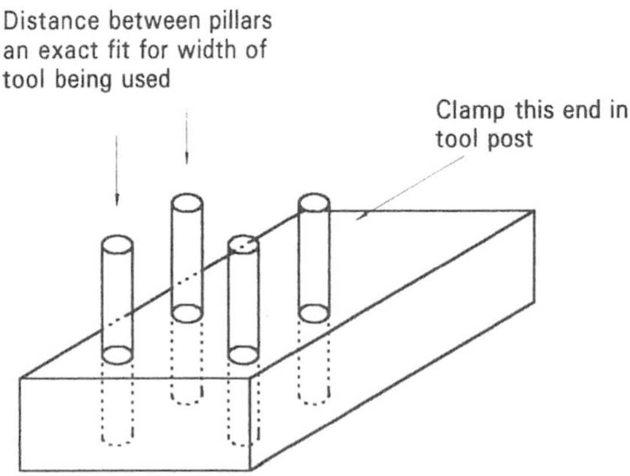

Device for supporting the tool when making a continuous groove in the fusee.

adjusted in depth as the carriage moves along. This tool is not difficult; make up a piece of steel with four pins fitted in it, as shown in the drawing (opposite), and clamp it in the tool post parallel to the lathe axis and fit a suitably radiused tool in the end of a square bar that slides nicely through the gap. Either fit a handle to the opposite end or wrap a quantity of insulating tape or similar material around it to avoid injury. (Do not use loose material; it must be held firmly on the tool with no danger of it coming off.) It is now possible to apply gentle hand pressure to the tool protruding through the tool post as the carriage moves along, and in this way to make the required grooves. If the first cut is not deep enough, the operation is easily repeated by picking up the groove with the tool before the lathe starts to rotate. The lathe must be set at its lowest possible speed for the operation, in back gear if possible. Better still, use a handle in the mandrel and rotate the lathe by hand. Although it is suggested that the tool be made from a square bar, this is purely to prevent it from rotating. Readers with sufficient confidence in their own ability can use a round one. The tool must remain at or near the same angle throughout operations to be successful. The operation is not difficult and there is a lot of enjoyment to be gained from the feeling of having created a nicely-shaped fusee. It is probable that the grooves will have a rough finish on them, so cut a piece

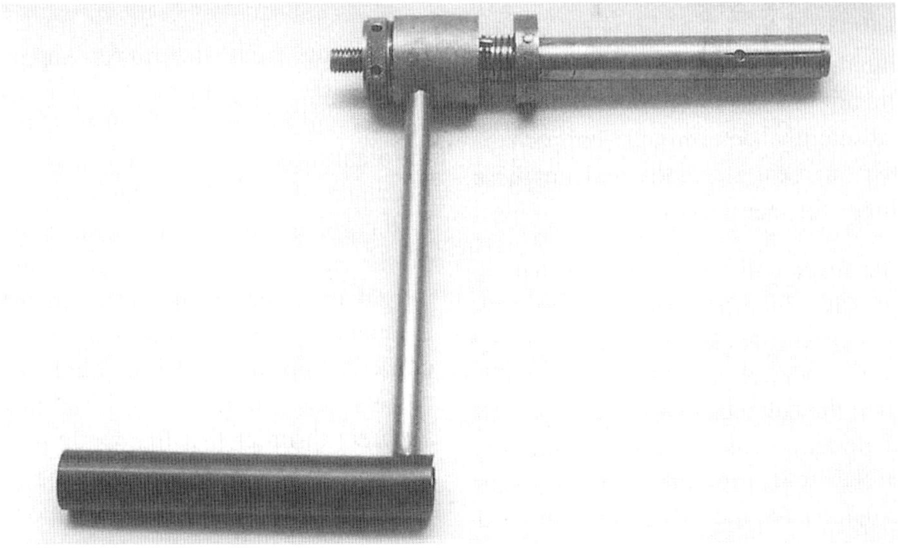

When making a fusee, the continuous groove can be a difficult proposition, and it is worthwhile to consider making a mandrel handle similar to the one shown.

of dowel to a shape that will fit them and, using Brasso or a similar polishing material, and with the lathe running under power and in back gear, run the dowel along the groove until a smooth finish is obtained. Finally, part the work off, or if it is thought to be too large to be parted off, saw it off. If sawn, mount it on a mandrel between centers, using a half center at the end that is sawn, which can then be faced to size.

Unfortunately the above methods only apply when a lathe has suitable screw-cutting facilities, and this is not always so. Generally it would be advisable for those without these facilities to purchase a fusee ready-made. Some people do not like to buy such items because they like the feeling of having made every part for themselves. For these people, it is possible with a little ingenuity to put on a continuous thread in these circumstances if sufficiently determined, or perhaps just to beat the odds and do these things for themselves.

The fusee will have to be finished, as far as shaping and drilling is concerned. Make a suitable well-fitting mandrel, but fit the end that will go in the tailstock center with a length of studding with a coarse thread. This thread will ultimately be the one transferred to the fusee, so it will need to have a large diameter in order to get a sufficiently coarse pitch. If studding is not available, it may be possible to obtain a large diameter bolt; a metal scrapyard is a good place to search for something suitable. The other end of this must be centered and supported by the tailstock. Take a nut that fits the thread and silver solder, or in some other fashion fit a short length of mild steel bar to it. If it is large enough, perhaps a couple of small screws would do the job. Use a piece of bar at right angles to this to connect to the screw-cutting tool via a slot in the first piece. When the lathe is rotated, the tool will now move along the thread and, with the tool in contact with the fuse, the continuous groove will be made. Most people with the small lathes that are the ones likely to lack screw-cutting facilities are unlikely to be making a clock that would require a large diameter fusee, and this method will therefore work quite well.

Continuing from the power supply toward the escapement are a pair of wheels and pinions designated as the third and hour wheels. They form the main part of the train of wheels known as the going train, which connects the power unit, whether it is a drum and weight or a spring and fusee, to the escapement. When first looking into a clock, it appears to be a jungle of wheels and pinions, and it is this apparent disorder that frequently puts people off making or repairing a clock. While these gears and pinions may be in a number of combinations, the actual formation is the same for virtually every clock. The great wheel drives a pinion, which in turn drives the center wheel; the pinion connected to that goes to the third wheel, and the pinion for that is in turn connected to the escape wheel.

Basically, therefore, we have four wheels and three pinions, which is a nice, easy, manageable number, particularly for anyone versed in engineering matters. Various combinations of wheels are used, but there must be a logical sequence. Let us start with a hypothetical clock, although the train used will be one that is quite common.

The great wheel connected to the drum arbor has ninety-six teeth and will connect with an eight-leaf pinion on the same arbor as the center wheel. The center wheel has sixty-four teeth and, as it carries the minute hand, must rotate once an hour. This in turn connects with an eight-leaf pinion on the third wheel arbor, which has sixty teeth. It connects to another eight-leaf pinion on the escape wheel. When designing a train, it is essential that the escape wheel make sixty revolutions (seconds) for each one revolution of the center wheel (minutes). To check this, multiply together the number of all the teeth in the driving wheels and divide the answer by the numbers of leaves in the pinions, multiplied together. With the example that has been used, the formula in the appendix gives all the details required to see how it works out.

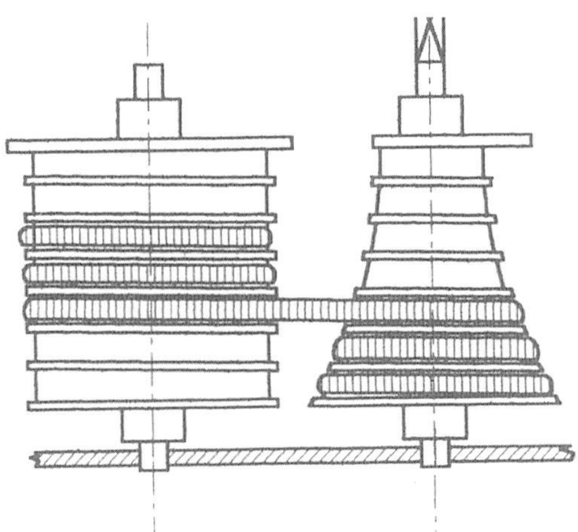

Drawing showing the principle of the fusee. The line can be gut or a chain can be used. At one time only a chain would have sufficed and it would have been handmade.

The assembled train showing the "'scape" wheel and pallets.

Chapter 4—Pendulums

A pendulum is described as a heavy particle suspended from a fixed point by a fine, inextensible, massless rigid thread so that it is free to oscillate on a vertical plane. Galileo (1564–1642), the famous astronomer, is credited with discovering it, and legend has it that he was watching the movement of a swinging lamp in church, which he timed by using his pulse. The important thing about it was that the oscillations were isochronous, which means simply that the time for one complete oscillation was always the same. His son is credited with the first use of the pendulum, and whatever the truth of the story, it has stood horologists in good stead ever since.

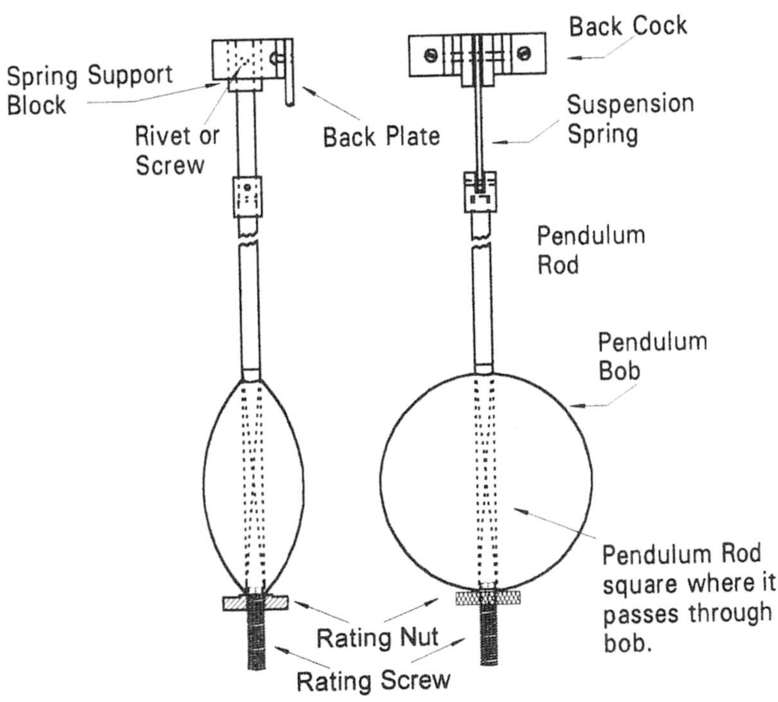

Makeup of a pendulum assembly, showing suspension spring and method of adjustment.

Although a pendulum appears to move backward and forward, it actually takes a sinusoidal motion—a movement that is not quite a circular one, but which for all practical purposes we can consider as an ordinary backward and forward movement. This is controlled by gravity, and as a result, the distance from the train to the bob varies according to where the pendulum is in use. There is also a variation in gravitational pull, depending on the height above sea level. The latter is such that adjustments can easily be made to take care of the variation.

Together with the escapement, and via the suspension and crutch, the pendulum is responsible for maintaining time, using Galileo's discovery that each oscillation takes the same period of time. A simple pendulum consists of three parts: the rod, the bob, and a rating nut, all requiring careful attention if the clock is to function properly and be accurate. The number of times the pendulum vibrates, or swings, depends on its length and is selected according to the type of clock being made. Long case and wall clocks more often than not will have a pendulum that vibrates once per second; smaller clocks will have

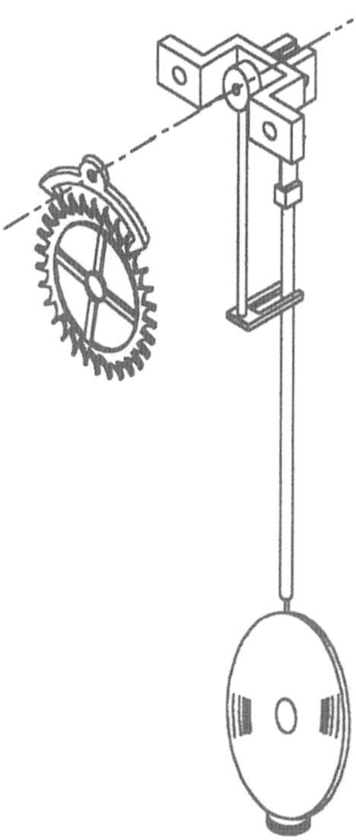

Relationship of pendulum to escape movement.

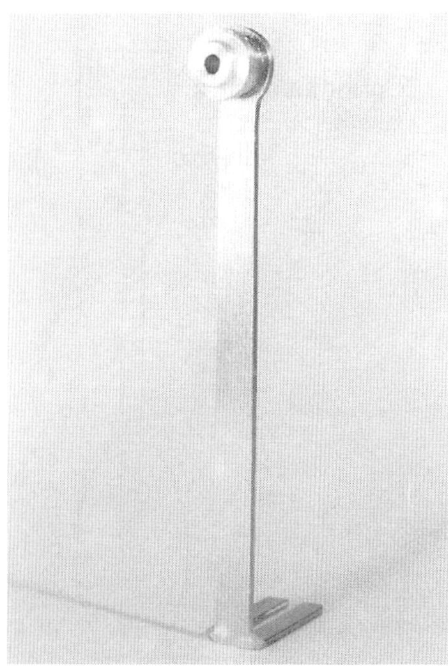

The crutch is the part that is moved by the swing of the pendulum. It fits on the escape arbor, the prongs on either side locating against the pendulum suspension spring.

correspondingly shorter pendulums and will vibrate faster.

This means that the length of the pendulum is also in direct relation to the clock train: a pendulum that vibrates every second will need a different train than a half-second one.

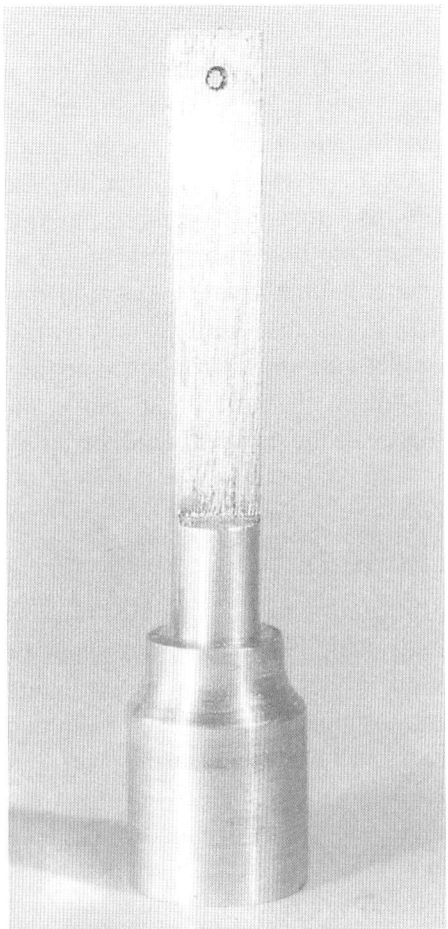

The pendulum rod will need two end pieces, one as shown will connect it to the spring suspension and must be slotted and drilled for this, the other will need a means of attaching the rod to the rating screw, and the method of making the latter will decide what the connector will look like.

Suspension

The suspension consists of a flat spring that is strengthened at the ends with metal blocks, allowing one end to be supported by the back cock and the other to connect to the pendulum. The spring must not be too long or too strong, and at the same time it must not be too weak, either. Most designs will give information on the correct size of spring to be used; if not it will have to be a matter of trial and error. Fortunately, there is a reasonable amount of latitude available in spring selection, but if in doubt, it is worth considering the purchase of a ready-made unit from a supplier, having in mind that unless the spring selection and makeup is right, the clock will not function properly. The back cock also needs to be made and assembled carefully, for if it is out of alignment, the pendulum cannot function properly. It must also be made with sufficient strength to give good support to the setup.

The Rod

The rod must be straight and have a suitable means of connecting it to the suspension unit at one end. The end that fits through the bob is threaded to accept the rating nut. Depending on the type of bob, it may be necessary to make the end square so that the bob cannot twist out of line. Selection of material for the rod is important. It has already been stressed that for correct operation the weight of the pendulum must be concentrated in the bob, and if a heavy rod is employed, this will take the weight away from that area. There is also a slight problem of expansion and contraction of the rod with temperature changes. Many ideas have been introduced to compensate for this, some simple, others very complicated. Whether or not the home clockmaker will want to go to the effort of making a compensating version is a matter of individual preference. Doing so involves considerably more work than making a simple one, but no doubt could in itself be an interesting project.

Assuming that, at this stage anyway, readers are going to be content with a single rod, then the material chosen for it is worth consideration. While brass looks nice, it is also the metal that suffers most from temperature changes, so if it is used for reasons of appearance, the rod should be as thin as possible, or alternatively use thin-walled tubing. Aluminum tubing is also a useful material and has the advantage of being light and allowing the weight to be concentrated in the bob. Wooden rods have the advantage that the material is less likely to expand or contract with climactic conditions, and on a long case clock, a piece of dowel can make a very effective rod, because it is also lightweight. It should be well sealed with varnish or a similar medium to prevent the absorption of moisture from the atmosphere. Make certain that the dowel is perfectly straight in the first place, as indeed must be any material that is used. A more modern material is Invar, which, although expensive, may not be prohibitively so, and has the advantages of being light, easy to work with, and not subject to changes of temperature.

The end of the rod has to be threaded to accept the rating nut, which allows adjustment of the height of the regulator. If a round or tubular bob is being used, then the rod can be left round where it passes through; if one of the flat types is used, then the rod will have to have a square on it to prevent the bob from twisting in use. Ensure that the square is positioned in such a way that the bob will remain parallel with the train when swinging. If allowed to twist at an angle, the balance of the pendulum will be upset.

The Bob

Although we tend to think of bobs as being made either in a lens shape or round, in fact, many clocks used quite fancy shapes; in particular French clocks were adorned in this way, but of course many French clocks were noted for their ornate appearance anyway. In general, the home clockmaker is going to use one of the two standard types, so it is these we will concentrate on.

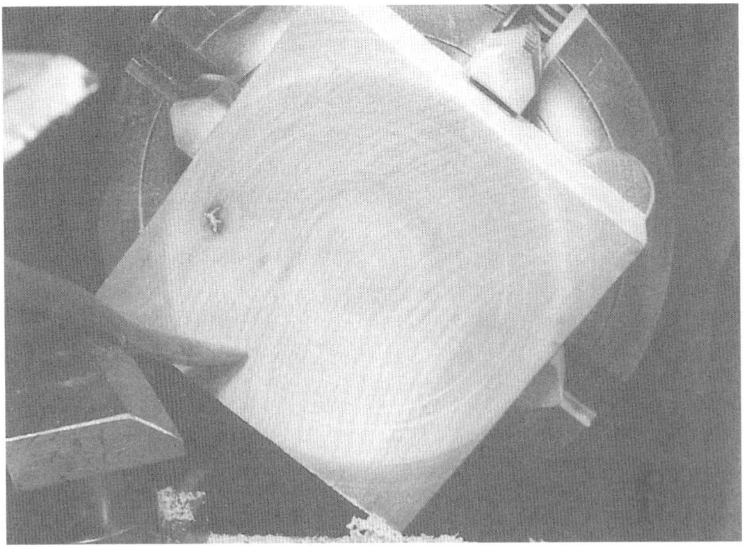

To make a lens-type pendulum bob, start by hollowing out a piece of hardwood to a suitable diameter and depth, ensuring the hollow has a nice smooth finish.

For small clocks and probably wall-mounted ones as well, the standard lens-shaped bob is usual and not difficult to make. Start by machining a hollow in a piece of hard wood; make sure the machined surface is perfectly smooth. Cut out two disks of brass a little larger that the outside diameter of the proposed bob and anneal them, if necessary re-marking the circles after annealing. A good alternative to brass is gilding metal, which is much more ductile and needs less annealing than normal brass sheet. Using a wooden bossing mallet, shape the two disks by gently hammering them into the hollows in the block. As soon as the brass starts to work-harden, anneal them again and keep doing so whenever necessary; do not under any circumstances try and work the metal if it hardens. During the shaping operations, take care that the work is kept central to the circumference of the hollows. It is all too easy to have the work slide to one side during the hollowing-out process, in which case the bob will be of no use at all. Periodically put the two pieces together to check the fit, until the edges meet all the way around. Each piece needs a small section filed out for the rod end, and at this stage little more than a nick will do. When satisfied with the hollowing-out process, drill a hole about 5⁄16" diameter in the center of one of the pieces and deburr the holes.

Clean the pieces up by soaking them in a solution of citric acid; two tablespoons to a bucket of water is about right. They need to soak in it for a

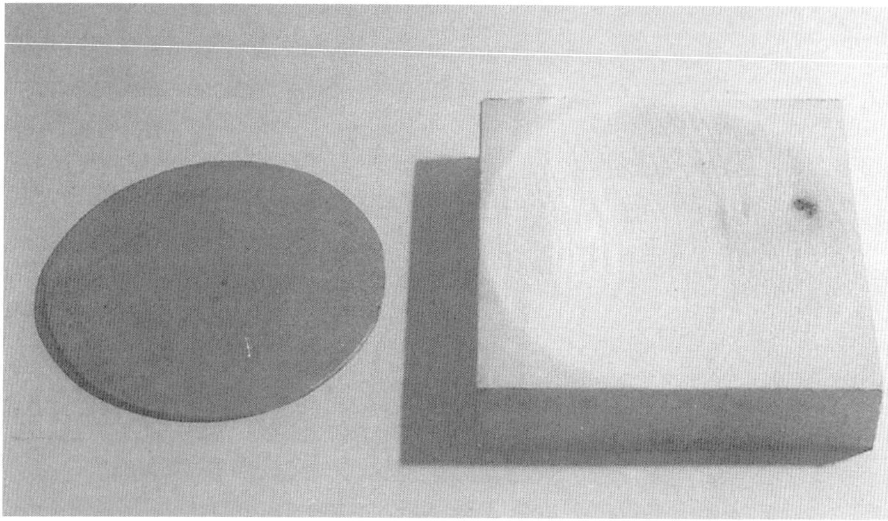

Cut a disk of brass to the diameter of the hollow and thoroughly anneal it.

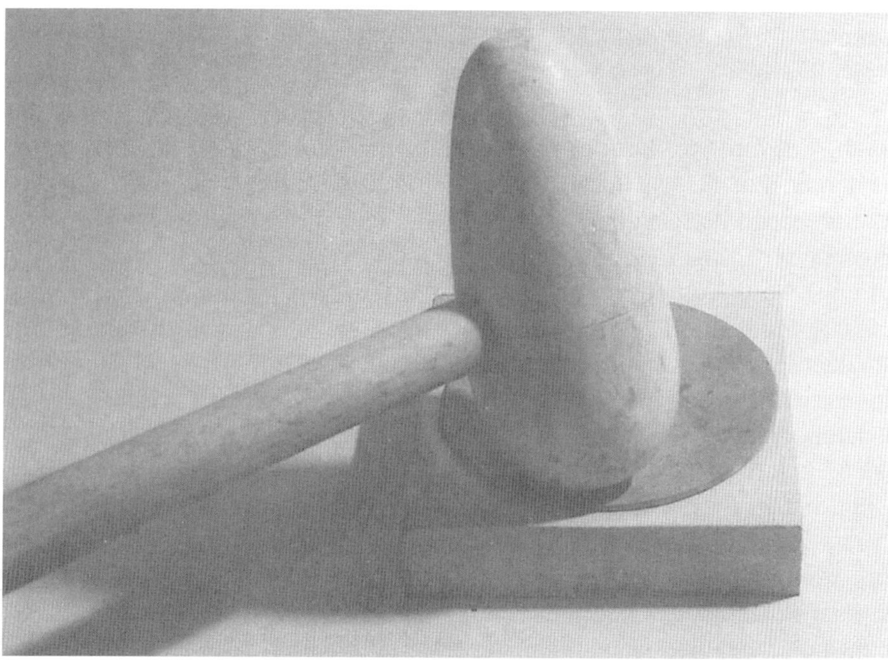

With a wooden bossing mallet, as used by car body repairers, shape the brass disk. Start by going around the outer edge and work gradually inward. Do not hit the work too hard or it will distort, and anneal it each time there is the slightest sign of the metalwork hardening. Make a second identical disk, and drill a hole for filling purposes in one of them.

couple of hours to ensure they are clean enough to be silver soldered together. Make up a solution of flux by mixing it with methylated spirits into a smooth paste and spread it around the inside edges of each piece. Put the two pieces together, with the piece with a hole in it on top, making sure the edges meet and that nicks that have been filed out are level with each other. Lay the assembly on a brick and put a weight on top so the pieces will retain their position. When completely satisfied that the parts are located correctly, heat them up and apply silver solder to about three or four places around the edges. Cool it off and put it back in the acid to clean it.

With a needle file, open the places where the nicks have been made and file a square for the pendulum rod to go through. The top hole can be opened with a taper reamer to obtain a round hole. Machine a length of brass tubing so that it is a push fit in the 5⁄16" diameter hole. Make up a short length of rod the same diameter as that used for the pendulum and with a square on it that

A good alternative to lead is one of the low-melting point alloys sold for casting in rubber molds. These also have the advantage that they are not toxic in the same way that lead can be.

Silver solder the two disks together, in three or four places, tap the filling hole, and fit a suitable piece of threaded tubing. Use this to grip the work in the lathe, and machine the edges of the bob so they are concentric.

is also the same. Paint this, using either emulsion paint or by using white-out. When dry, push it into the bob, fit the tubing into the hole, and then heat some lead to melting point and pour it through

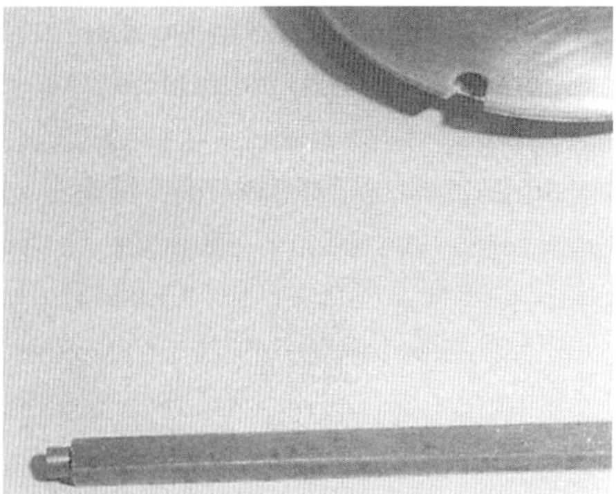

The round pendulum rod will fit in one end and the other will need a square hole to prevent the rating screw from turning when adjustments are made. File a square in one end of the bob for this and exactly opposite a smaller round one. Make a small round section on one end of a square rod and push it into the bob. Check that it is a nice tight fit and that the square edges are not angled to the bob. When satisfied with the fit, remove it and paint it with emulsion paint. When that is dry, cover it with grease and push it home in the bob. The paint and grease will ensure it can be removed after filling.

the tube until the bob is full. It will spill out of the edges but this is of no consequence because any surplus can be cleaned off later. Solder a disk in the 5⁄16" diameter hole in place of the tube. Finally, use a fine file, emery paper, or cloth, and any other abrasive material that might be a personal favorite, to tidy the surfaces of the bob and make it look presentable.

Round Bobs

Round bobs at first glance seem to be far easier to make than the lens type, and in some ways they are; this does not mean that care need not be taken in their manufacture. The finish that is required has considerable bearing on how they are made. For example, some people are content to drill a piece of cast iron bar and use that, which when cleaned and painted can look quite good, while at the same time it is the minimum of

The assembly looks rather like the above photograph. Fill the bob with lead or a low-melting-point alloy. To make sure it is absolutely full, keep a small flame underneath it while filling so the lead cannot set and prevent a full flow through. When the bob is full, allow it to cool and remove any lead that has spilled through the gaps in the edges. Take out the tubing that has been acting as a funnel (it may need the application of a little heat) and seal the hole with a brass plug, soft soldered in place. Finally give the whole thing a good clean up with files and abrasives, removing any uneven sections, then polish and take out the rod.

trouble to make. The only thing that needs particular care and attention is to ensure that the hole drilled should be a nice fit for the thread for the rating nut. The same principle could be applied using a length of brass rod, and this would give a better appearance but less weight.

A more common way is to use a piece of brass tube, and after ensuring both ends are perfectly square, start by silver soldering a piece of plate to one end. This is then trimmed to the edges of the tube to give a nice neat finish with a hole drilled centrally in it. The screw for the rating nut will be traveling through a piece of brass tube fitted inside the bob, and the hole should be a close fit for that to slide into. Make up a length of mild steel bar with a step to fit the inside of the other end of the tube and a hole drilled centrally in it to accept the central tube, referred to above. Make sure that this tube is exactly central when fitted, otherwise the bob will be off center and this will interfere with the operation of the pendulum. Fill the outer tube with lead. If necessary, heat the base to ensure that the material fills all the space. A good alternative to lead is one of the low-melting-point alloys sold for casting in rubber molds. These also have the advantage of being not toxic in the same way that lead can be. Depending on the amount of weight required, the tube does not have to be filled to the very top, but make sure that the filling material fills the area completely and there are no air pockets that could throw the bob off balance. The final act is to remove the steel spacer and put a brass plate over the top; this can be soft soldered in position. Alternatively, a brass end can be made with a lip so that it is a push-fit in the tube. This allows more weight to be added if required.

American and French clocks often use a fancy-shaped bob and these are usually cast; they can be bought. To make them oneself requires either a high degree of skill in beating sheet metal to shape or the ability to make suitable patterns to have the shells cast. Any casting would be best done using the lost wax process. This would mean the bob coming straight from the foundry with a finish suitable immediately for use. Trying to clean up an ordinary sand casting would be far from easy, and even then it is doubtful that a suitable finish could be obtained.

The Rating Nut

The rating nut is made to fit the thread on the rod and can take almost any form or shape that the clockmaker likes, although it is as well to ensure that it can be adjusted by hand easily. Fit the rod through the bob and put on the nut to check whether or not the rod will move as the nut is adjusted. If it should stick at any point, ease the inside by driving a piece of square mild steel through the hole, using a hammer and tapping lightly. The steel will be sufficient to clear any lead that might just be in the way. This should not apply to tubular pendulums with a center tube.

Chapter 5—Escapements

The escapement is part of the mechanism of a clock that is devoted to the speed at which the clock works. It consists of a coarse, specially shaped toothed wheel and an oscillating bracket with two pegs, called pallets, that locate in the teeth. The name "escapement" is used because this is the point where the power of the clock, which is derived from a weight or a spring, is prevented from escaping and is converted from rotary to reciprocating motion. The arbor on which the escape wheel, or as it is often called, "'scape wheel," is located is connected via a pinion to the main wheel train. The action is such that a tooth of the 'scape wheel is trapped and then released at regulated intervals, and this movement in turn is sent back through the train to the hands. Without

This demonstration model of a foliot escapement is part of a collection of such examples held by the British Horological Institute.

the escapement, the train would just unwind at no given speed and the clock would be of no use. Many escapements work at a rate of one movement per second, so if a hand is fitted to the end of the pivot on which it runs, the seconds can be counted off.

There are many types of escapement—so many in fact that whole books have been written on that subject alone. The beginner to clockmaking is generally only likely to use one of about four types, but anyway, any clock plan one might work from is certain to give full details of construction. Even so, it is useful to know what we are aiming for and how to set about making this part of the clock.

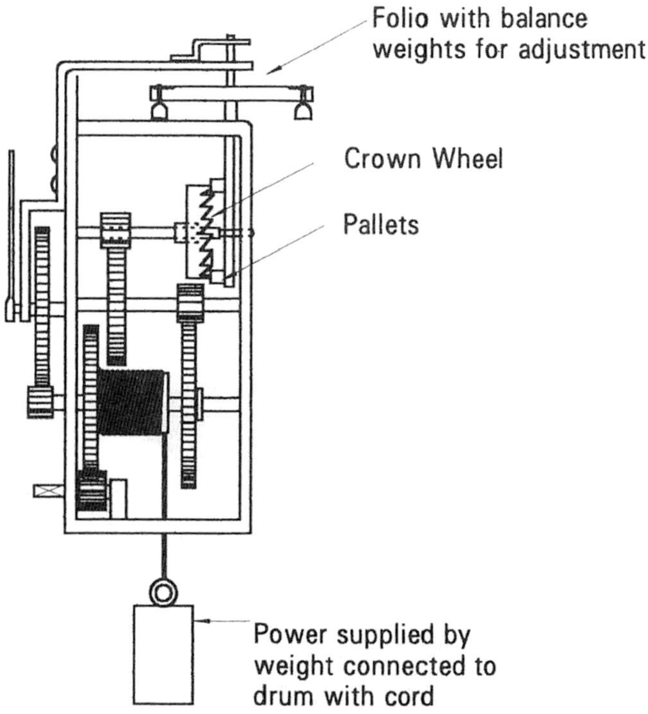

Sketch showing the use of a foliot and verge escapement.

The Verge and Foliot

The earliest form of escapement known to be in general use was the verge and foliot.

It is a very simple device to make, but unfortunately it is not known for good timekeeping. Even so, it is an interesting experiment to make one, because it gives some idea of what we will be looking for when making a more advanced version. The 'scape is in the form of a crown wheel, with the teeth at ninety degrees to the movement. There is a slight radius on each tooth, and the number of teeth will depend on the numbers used in the train.

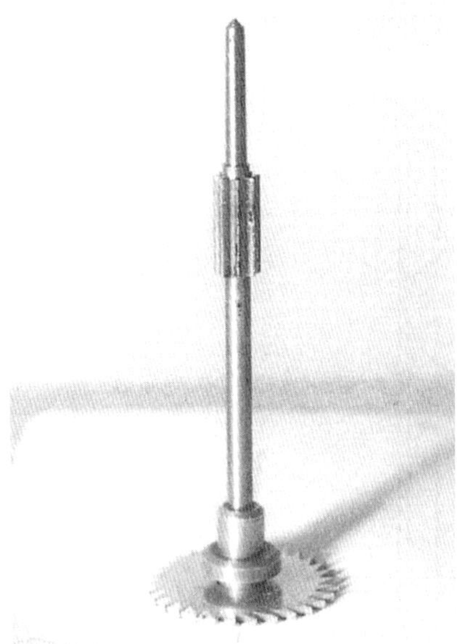

Escapements are usually marked on paper and then transferred to the metal for the pallets to be shaped. This requires the 'scape wheel pivot to be set at right angles to the paper and can be achieved by making a small hole of the right diameter in a piece of wood.

The movement of the 'scape wheel is regulated by the pallets, which are pieces of basically flat metal, shaped to a knife edge, attached to the verge that runs vertical and is positioned so that the pallets enter the teeth of the 'scape wheel. These pallets are set at an angle of just over ninety degrees to each other, and as one tooth of the 'scape wheel passes one of the pallets, it is locked by the one opposite. The shape of this causes the locked pallet to be pushed out of the way and causes the foliot, which is a cross bar, to swing, and in doing so it locks the tooth on the opposite side. In this way the process is repeated continually, with the foliot moving backward and forward as each tooth is locked and unlocked.

The speed at which the mechanism operates is adjusted by weights at each end of the foliot, in which a series of grooves are cut as a means of holding the weights in position. Of course, the foliot must be well balanced, and the grooves spaced evenly on either side, so that a good balance is maintained. It is a simple idea and, as such, served clockmakers well for many years until more sophisticated ideas came along. The nearer the weights are to the center, the faster the movement will go, and vice versa; if the weights are taken toward the end, it will slow down.

To make the escapement, the wheel can be made on a flat plane and then rolled into circular form and the ends silver soldered together, then a cross bar in which a hole for the pivot has been drilled is silver soldered on. Needless to say, it is essential that the pivot hole is placed central to the circumference of the wheel, and a brass bush will also need to be fitted to hold the pivot. The teeth can be machined to shape, but this is an example of an escapement where it is possible to carefully file the teeth to shape with sufficient accuracy for the device to work well enough. As a means of getting an idea of what clockmaking is about, the escapement has much to recommend it: connect to a couple of wheels to give sixty-to-one reduction, fit a winding drum and weight to that, and we have a crude clock which will show seconds and minutes. Elsewhere, details will be found of how to convert this to hours, and so using little material or time a clock can be made in this way. It will not be all that accurate, but it certainly makes an interesting starter for anyone who feels that clockmaking is a difficult art.

A later development of the verge and foliot escapement was the verge and balance, the foliot being replaced by a wheel, and speed was altered by either changing the driving weight or by changing the depth to which the pallets entered the wheel. Unless a replica of a particular clock that used the device is being made, it is not something that is likely to appeal in general to the amateur.

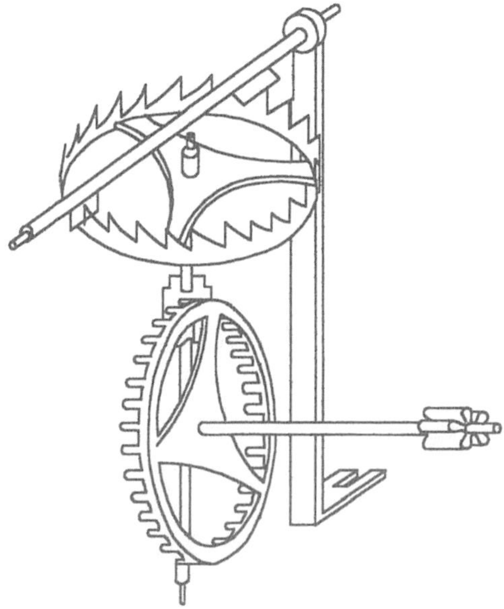

Verge adapted for use with pendulum.

Verge and Pendulum

It is possible to use the crown wheel and verge and, while retaining its simplicity, improve accuracy considerably. The escapement works in exactly the same way, except that the foliot is replaced with a pendulum. This is made with a crutch in exactly the same way as pendulums used with other escapements, and full details of how to go about those will be found in the relevant chapter. One major difference in the use of the idea is that the 'scape wheel will now lie horizontal, so the direction of movement in the train will need to be changed. In normal engineering practice, we would use a pair of bevel gears for such a purpose, but in clockmaking the method is to use another crown wheel, set in the vertical position and locating with the pinion attached to the 'scape wheel.

Recoil Escapement

We now come to the more common type of escapements that will be found in many of the published designs, and the first noticeable difference is that the teeth are cut in the edge of the wheel rather than at ninety degrees as before. The pallets are much more compact and work from a pivot situated above that of the 'scape wheel, which generally has thirty teeth, but this will depend on the train. The action is similar to before;

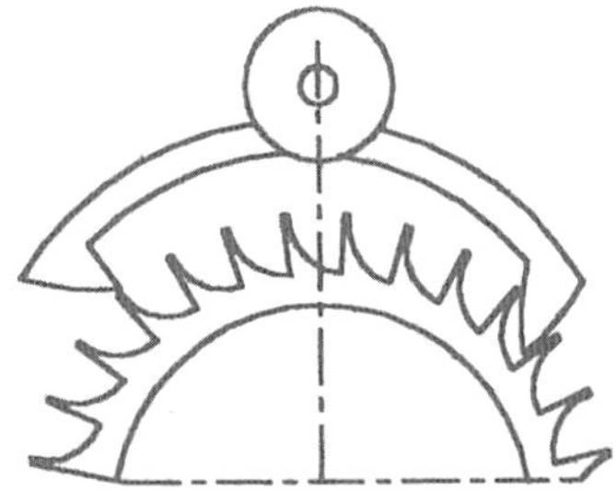

Recoil escapement

The pallets for a recoil escapement, made by marking on paper and sticking the drawing to the metal.

one nib of the pallet locks into a tooth and then, as it is unlocked, provides a slight impulse as the result of the shape of both teeth and pallets, while the other in the meantime is locking a tooth farther along. The movement is such that there is a very slight backward movement as unlocking takes place, an action known as recoil.

Generally, full details will be given on any plans of how to lay out the escapement, which must be planned as a whole in order to find the correct proportions. The pallet centers are recommended to be at a distance of 1.4 times the wheel radius from its center, and although there are occasions when this may vary, we will use it is our figure. All that is needed then is to draw a centerline, mark the position of the wheel center, measure 1.4 times the radius, make another mark, and we have the correct position. Sometimes that figure of 1.4 can be an extremely awkward one to measure, and if so, the position can be obtained by the use of some simple geometry. Start by drawing the circumference of the wheel and, from the center point, draw two lines at forty-five degrees. Where these intersect the circumference, draw lines at right angles toward the centerline. The position at which these two lines cross is the one where we want the pallet arbor to be.

Assuming the 'scape wheel will have thirty teeth, their positions can be marked on the drawing. It is not necessary to mark the position of all the teeth, about ten will do. There will be twelve degrees between each, but an allowance has to be made for the fact that there is a flat on each one and it is usual to allow one degree for that. Use a protractor to put the positions on the drawing. Readers who have a computer with Computer Aided Design (CAD) software will find that drawing the escapement is very easy indeed; using a protractor and ruler with a pencil is

Escape wheels: one made by using a commercial cutter, the other by filing the teeth to shape.

more difficult. An allowance of one degree is also left for the drop, and this means the pallets are ten degrees of the circle.

Once the pallets have been drawn, it is customary to cut out the drawings and stick it to a piece of gauge plate, and to cut and file around it to get the required shape. To an engineer it may sound a rather primitive way of going about things, but the system has worked for clockmakers for hundreds of years and there is no reason for anything more sophisticated.

Dead Beat Escapement

Unlike the previous escapement, in this case, when the pallets lock on to the 'scape wheel teeth, there is no recoil action, hence the name "dead beat." It is a comparatively easy device to make and is capable of producing accurate timekeeping, although it should be pointed out that it is really more suitable for large clocks than small ones. George Graham (1673–1751) invented it around 1730, and so it has stood the test of time. It is particularly effective with weight-driven clocks where there is a constant source of power. Generally, the 'scape wheel will consist of thirty teeth and the pallets span from eight to fourteen teeth. The pallets are relatively easy to make and the teeth can be cut with a fly cutter. As with all escapements, the pallets should be hardened and polished.

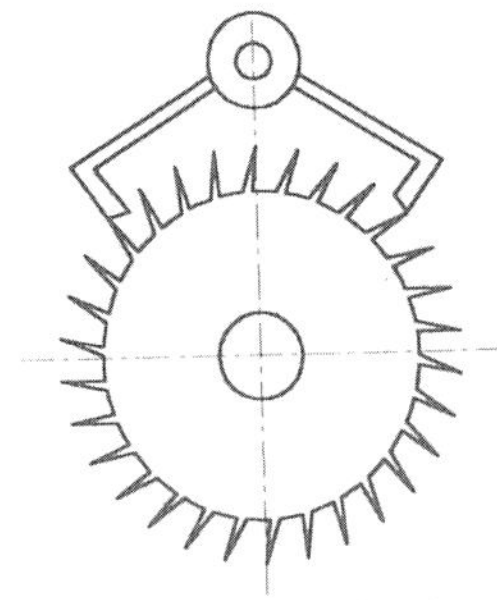

Dead beat escapement. Note that the teeth are undercut by six degrees in order that only the tip will be in contact with the pallets.

Brocot Pin Pallet Escapement

Those readers who wish to go it alone and design their own movements might well be interested in the brocot pin pallet escapement, the escape wheel for which is virtually identical to that for the dead beat. The pallets, however, are completely different. They are in pairs instead of the more normal nibs that are found in the previous two escapements. Half-round sections are used, and these can be made from round drill rod,

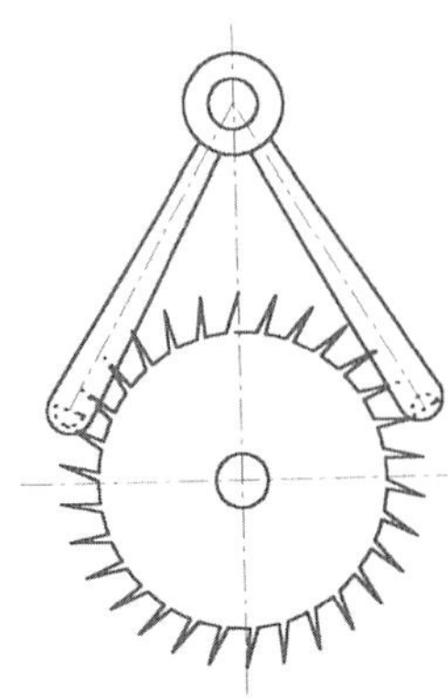

Brocot pin pallet escapement

stepped to fit in holes in the arms and filed or milled exactly in half. Only these parts need to be hardened, and the arms can be made from mild steel and the nibs secured with a suitable retaining compound.

Pin Wheel Escapement

Although at first glance this type of escapement would seem to be the simplest of all to make, since there is no escape wheel as such to cut teeth on, it is not quite as straightforward as it may seem. Consisting of a wheel with a series of holes into which are inserted pieces of half-round steel, in a similar fashion to the nibs in the brocot escapement, the spacing of the holes is a simple enough proposition, but it is essential to ensure that when the pins, which are usually made from a good quality brass, are halved, this is done with absolute accuracy. It is therefore necessary to do the work in a milling machine or by using a vertical slide on the lathe to obtain the required accuracy. It will involve making a jig that will hold the pins securely in such a way that the cutter can reach the centerline of the pin. Mounting the pins accurately on the wheel also requires a simple jig to be made up. This can be from a stiff card, if one wishes and assuming it is only to be used once, otherwise mild steel should be used. The arms are made in two parts and are fairly straightforward and will present little problem to anyone even slightly versed in metalwork techniques.

Pin wheel escapement

French Pin Wheel Escapement

This, too, is an escapement that might prove of interest to the model engineer wishing to go it alone and design his or her own clock, because it is straightforward. Unlike the brocot escapement, the pins are left round and the arms of the pallets fit either side of the wheel. It is very attractive and elegant to see when working. Because the pins are round, they are easier to make, and if a modern material such as Teflon is used for the pallet nibs, the problem of lubrication, which at one time made the escapement unpopular, need be no problem at all. Both types of pin pallet escapement require the pallets to be offset from the wheel, rather than in line, as is the case with the previous types.

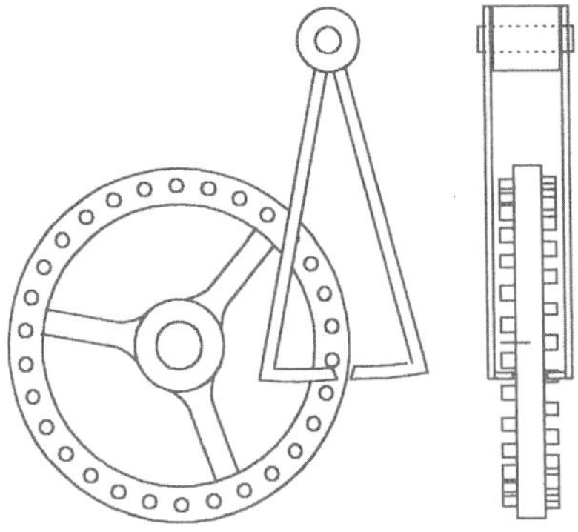

French pin wheel escapement

Making Escape Wheels

The wheels of any type of escapement are always more difficult to make than the normal wheel found in the train because of their shape. Special cutters can be purchased that are designed to deal with a particular type of wheel, and so are only usable for the one escapement type, and reference to the various drawings will show readers why this is so. Suitable cutters can be made for the recoil type of escapement from drill rod and, as only one radius and a straight edge is required, they are easy to make. The cutter can be made as a fly cutter but it is far better to use a multi-toothed cutter if possible, because a fly cutter always seems to apply too much force for comfort as it makes its single cut on each rotation and there is always a tendency to feed a fly cutter into the work too fast. To make a six-tooth cutter is not a great deal more difficult than making a fly cutter and it will be much easier to use.

First of all, a form cutter will be required in order to get the required radius; use a piece of ground-flat stock, or gauge plate, as it is more often referred to these days, and drill a hole using a drill with a diameter twice the radius needed. File or saw off the metal left at the side of the hole and just file a slight relief on the front edge that is left; do not touch the radius that remains. A slight chamfer can also be put on the top if one wishes, but this is not necessary, as the tool is only going to be used for the one job. Cut the flat stock to the required length and soak it in water and dish soap. Without wiping the liquid off, heat it until it is a bright red color, similar to the color of a boiled carrot, and then quench it as quickly as possible. This can be done with saltwater, but then there is a slight chance of the metal cracking as it cools. It is far better to use a vegetable oil of some sort or another. There must be sufficient quantity of the oil to ensure a thorough cooling; it is no use trying to quench the metal in a cupful. It is well worth investing in a bottle of cooking oil and keeping it in a suitable container just for this purpose. Even if, so far, you have never had cause to harden metal, there will be two or three instances where you will need to when clockmaking, and it is as well to have some available. Take care: sometimes the oil will catch fire. If so, do not try and put it out with water, but just exclude the air. Any such happening is unlikely to be a fierce blaze, but work on a suitable surface, just in case hot oil splashes out. The best type of container to use for the purpose is something like a cookie tin—the lid will not only keep the oil in place when not in use, but will also put out any flames that might result from quenching the metal.

Clean off the scale that has appeared on the metal, and when it is nice and bright, place it in a small can (a sardine can will do nicely) that is half-full of sand. Heat the sand from underneath until the metal turns a dark straw color and then quench it in the oil once more. Finally, just rub the top of the cutting

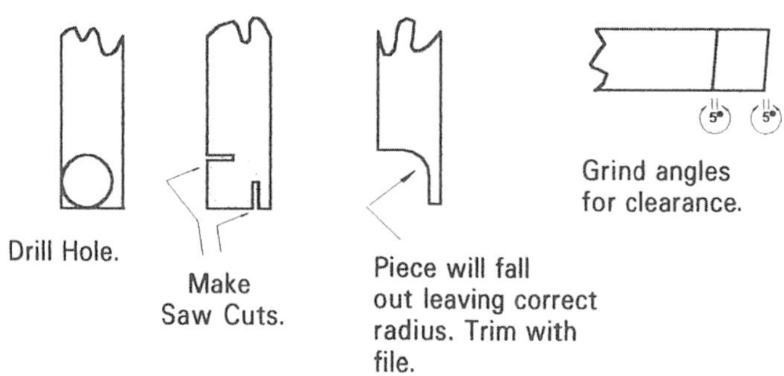

Making a form tool to produce a multi-tooth escapement cutter.

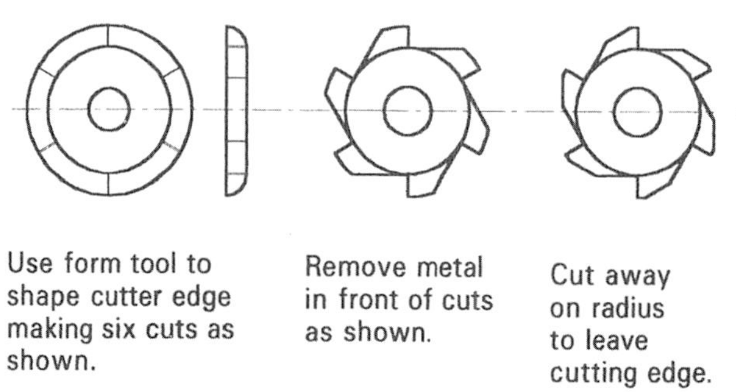

Method for making a multi-tooth cutter. Use drill rod. The same principle applies to wheel and pinion cutters.

edges with a small oilstone to put something of an edge on them, and the tool is ready for use.

To make the actual cutter, start with a length of drill rod bar mounted in the three-jaw chuck. The diameter of the bar, within reason, is not important as long as when the mandrel to which it will be fitted is allowed for, sufficient depth is left to obtain the full depth of the teeth. The chart shows the radius and therefore the depth required for various wheel sizes. Machine a little off the outside diameter, just sufficient to take care of any eccentricity that the chuck has, and then drill the hole for the bore; we now know that the bore is perfectly concentric with the outside diameter. Gently run the radius tool you have just made until the edge of the drill rod bar has the required shape. Use a cutting oil for the operation, as the original form tool is only just about efficient enough for the job and needs any help it can get in rounding off the drill rod bar.

The final task is to part the bar off, and this is something which some people find difficult. Much depends on the quality and size of the lathe as to how easy it is, and if possible it is better to use a rear tool post for the work. For those who really cannot face up to the idea of trying to part off a piece of drill rod of this sort of diameter, there is another way around things. When the material is first put in the chuck, leave an overhang of about an 1½", or 40 mm, and still machine the outside edges for concentricity, but don't drill a hole. Turn the radius as described and then machine a stem at the back of the tool; assuming the work has remained in the chuck the whole time, the stem will be perfectly concentric with the tool. The bar can then be removed from the chuck and the part sawn off, and we have our shaped metal but with a spigot instead of a hole for mounting it on a mandrel.

To make the teeth, mark off or index it at six equal divisions and mill a flat as shown in the drawing (opposite). Finally, machine another flat so that there is plenty of clearance; again the drawings show what is required. We should now have six sections evenly spaced and the tool will work like this, but if the square edges have a small relief filed or machined on them, it will work even better. If, however, you are

Sizes for Making Cutters for Recoil Escape Wheels

Escape Wheel Pitch less thickness of tooth tip	Radius of Cutter
0.05"	**0.125"**
0.06"	**0.16"**
0.08"	**0.2"**
0.10"	**0.25"**
0.125"	**0.3125"**
0.16"	**0.40"**

not entirely confident of being able to get these relief angles without damaging what will be the cutting edges, leave things as they are. Finally, repeat the hardening and tempering exercise as detailed already and you have a completed 'scape wheel cutter. It will not be as efficient as a professionally made one, those have sixty teeth as a rule, but it will do the job, and after all it is unlikely to be used for cutting more than one or two wheels.

When cutting wheels and in particular this type, it is essential that they are well supported to as near the point where the teeth are being cut as possible, particularly where homemade cutters of any sort are in use. The pressure required to cut the teeth, even though we are only removing a little material at a time, is considerable, and the tool will tend to bend the metal as it cuts. That is the last thing we need. Generally, 'scape wheels are made from material of 1⁄16" thick and we are torn between two possibilities here. To get the maximum efficiency, the 'scape wheels need to be as light as possible, but they also do a great deal of work with the pallets banging up and down on them for twenty-four hours a day. While therefore it may be desirable to use something a little thinner to save weight, doing so could defeat its own purpose, as it is more likely to distort during operations.

Some readers may not have sufficient equipment to make the tools or not feel entirely confident about so doing. It is possible to file the teeth by hand after very careful marking out. It goes without saying that a great deal of care is needed, and in particular it is essential that the straight sections are really that and are not angled in any way. If a small template is made first, it can be used to check that the radius of each tooth is correct; it is almost impossible to do so when just working to markings. The template can be made from a piece of cardstock (card), but something more permanent is preferable and plastic is ideal. A suitable piece can be obtained by cutting up an old credit card and finishing the radius with an emery board of the type used for manicuring. On many very old clocks all the teeth on the wheels are hand cut, and in comparison to a great wheel an escape wheel is very simple indeed.

To complete the escapement we require the pallets, and to find the size and shape of these it is necessary first to draw the necessary angles. The pallets can then be cut out, using gauge plate to make them. There is really no way to make them except with a file and saw, although perhaps roughing out could be done on a milling machine. It is essential that the pallets are made accurately and that the nibs, which are the pointed pieces that stick out, have polished working surfaces. As much polishing as possible should be done before they are hardened, but in doing so the shape and size must be maintained. Final polishing can be completed after hardening; methods of so doing will be discussed elsewhere.

The final task to complete the escapement will be to cross out the wheel. When that is completed, mount it on a collet and secure it to the pivot. It can then be tested, using a depthing tool, to check that the pallets will do their job. As they are moved backward and forward, the action should be sufficient to slowly rotate the wheel.

Chapter 6—The Going Train and Motion Work

The going train, as it is properly called, connects the power unit, whether it is a drum and weight or a spring and fusee, to the escapement via a series of wheels and pinions. When first looking into a clock it appears to be a jungle of wheels and pinions, and it is this apparent disorder that frequently puts people off making or repairing a clock. While these gears and pinions may be in a number of combinations, the formation is the same for virtually every clock, and in fact there are nowhere near as many as it seems at first glance. The great wheel, which has already been discussed when dealing with the power, drives a pinion that in turn drives the center wheel. The pinion connected to that goes to a third wheel, and the pinion for that is in turn connected to the escape wheel.

We have therefore just four wheels and three pinions, which is an easy manageable number, particularly for anyone versed in engineering matters. The wheel combination will vary in size, depending on the size of the clock being made and on the whim of the designer, but must conform to certain parameters in order to maintain time. Various combinations of wheels are used, but these must be in a logical sequence. Let us start with a hypothetical clock, although the train used will be one that is quite common.

The great wheel connected to the drum arbor has ninety-six teeth and will connect with an eight-leaf pinion on the same arbor as the center wheel. The center wheel has sixty-four teeth, and as it carries the minute hand must rotate once. This in turn connects with an eight-leaf pinion on the third wheel arbor, the wheel for which has sixty teeth. It connects to another eight-leaf pinion on the escape wheel. When designing a train, it is essential that the escape wheel make sixty revolutions (seconds) for each one revolution of the center wheel (minutes). To check this, multiply together the number of all the teeth in the driving wheels and divide the answer by the number of leaves in the pinions, multiplied together. With the example that is shown in the Appendix on page 125, this works out as sixty, which is what we want; any combination can be checked in this way to ensure that the train is going to be right. In addition to this it is possible to work out the size of barrel and length of driving cord that will be required.

If a shorter pendulum is used, it will beat at less than a second, and from the table it is possible to work out at what speed a given length will beat. Suppose, therefore, the major factor behind the design of a clock is not as usual to fit a case around the movement but to make a movement to fit a case that one particularly wants. Reference to the tables (see Appendix, page 126) can enable us to find out the speed at which a pendulum of a given length will beat and what wheel combination is needed to make the clock work accurately. Although a table of common wheel trains is included (see Appendix, page 128), it does not mean that they are the only combinations available; it is quite possible to calculate one for oneself.

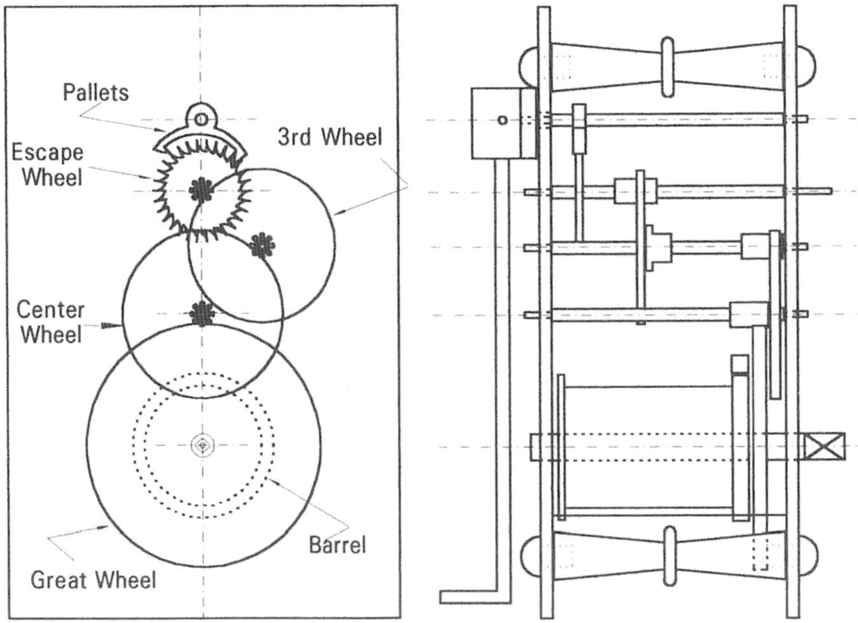

General arrangement of a clock. There may be some variation on position of pinions, which will mean adjustment of wheels to match.

Motion Work

The train then sets the clock so that it ticks off the minutes, but more than that is needed: every clock must show the hours, and some even include a calendar, while others have phases of the moon. It is not intended to deal with these latter factors or striking mechanisms, but it is necessary to arrange for the clock to indicate the passing hours as well as minutes.

As the minute pivot does one complete revolution once every sixty minutes, what is needed is a twelve-to-one reduction, in order to mark off the hours. This is obtained in the usual way by using a pinion and wheel and therefore any pinion and wheel with a multiplication of twelve will do. Another factor comes into it as well, and that is the size that the hour wheel will be; for example, with a six-leaf pinion the hour wheel will have seventy-two teeth, which is quite manageable. Make the pinion twelve leaves and the hour wheel needs a hundred and forty-four teeth, a size which is likely to be much too large for most clocks. Probably the largest size that can be coped with in most movements will be an eight-leaf pinion and a ninety-six-tooth wheel, and even that is on the large size; this leaves a limit of six- or seven-tooth pinions as practical propositions, because above that, things will become far too unwieldy.

So far so good; we have no doubt made up our minds of the best combination for the train, and it now becomes a case of connecting these two to the pivot that will operate the hour hand. The obvious answer would seem to be to put a suitable pinion on the arbor of the minute wheel, connecting it with a wheel that will give a twelve-to-one reduction on another arbor immediately below or at the side, and put the hour hand on that. There is only one snag: if we have a pinion or a wheel rotating in a clockwise direction and connect another directly to that, the addition one will rotate in a counterclockwise direction, which is not a great deal of use. Therefore, two additional arbors with suitable wheels and pinions are needed, the first to change rotation to a counterclockwise direction and the second to change it back again. In the meantime, any necessary reduction can be incorporated.

Some old clocks, and no doubt some being made at present as well, have the minute hand immediately below the hour one. These do not have the going train arranged in a straight line, as is the more usual arrangement. Additional arbors with pinions and a reduction wheel are then set at an angle to enable the hour arbor to be placed in the center of the movement. The finished result is quite attractive and worth considering.

Because space is generally limited, most clocks have the hour and minute wheel in the same place; tradition also probably has a part to play in the arrangement. The same necessity to change the direction or rotation is still present, and the way this is done is to fit a wheel with a suitable number of teeth on the minute pivot and to arrange an identical wheel adjacent to it with a pinion attached to it so it will rotate at the same time. As the wheel mates directly with one fixed to the minute pivot, it will rotate in the reverse direction. There is no arbor for this pair to run on, so a short one is provided and they are allowed to run free on it. It has a threaded end and screws into the front plate; frequently the other end is supported by a bracket with a hole to act as a bearing surface. In other instances there may be just a hole in the end of the pivot, through which passes a split pin to prevent the wheel and pivot from coming off. Bearing surfaces, of course, should always be of different metals, and in this case we have a slight anomaly as the reverse wheel is generally made of brass and will be running with a brass wheel, while at the same time being connected to a hardened steel pinion, on a pivot that is generally made of brass. The system seems to have stood the test of time, but could be improved with the use of modern materials. But then clockmaking is a traditional industry, which is one reason for its fascination.

With the pinion in place, it is now a matter of connecting it to the hour wheel, which is done by simply allowing the wheel to run loosely on an extended collet fitted to the minute wheel. The hour wheel in turn is fitted to a collet that is also a bearing of extended length to ensure it runs true and that there is sufficient surface to allow for wear. The collet on the minute wheel has a square on the end and the hand will be a push-fit on this. The hour hand can then be fitted directly to the collet that supports the hour wheel. Both hands are now running from a central point.

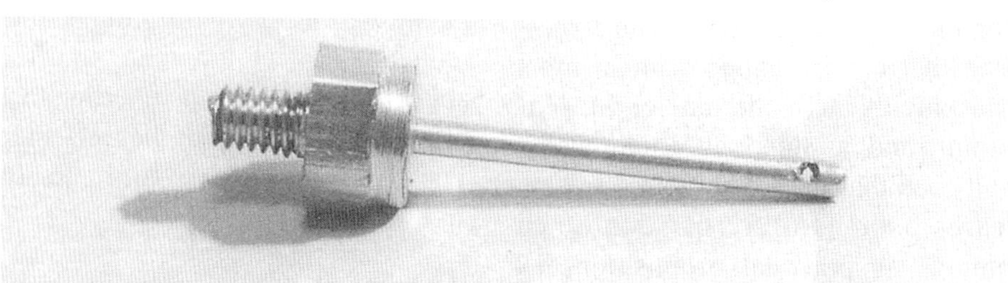

The small extra arbor required for the motion work that screws into the front plate of the clock. It takes one of the rotational change wheels and the hour wheel. In this case the arbor has been drilled to accept a pin; in other instances a bracket will be used for security.

One of the identical wheels used in the motion work to change the direction of rotation. It is fitted with a hollow pinion known as a cannon pinion, which will in turn line up with the hour wheel.

The pair of motion-work wheels awaiting final assembly.

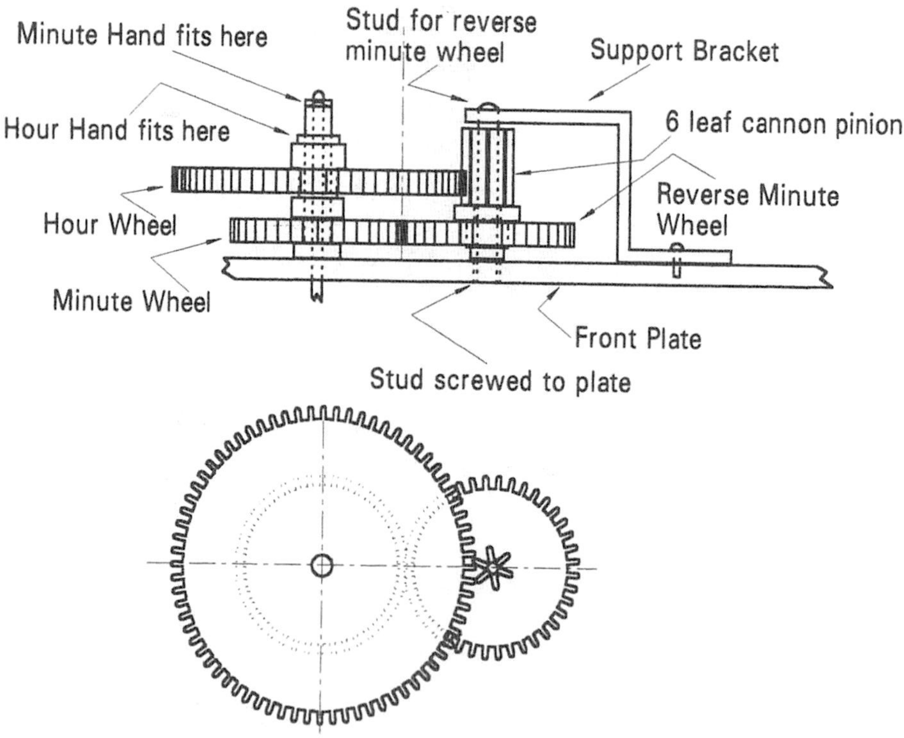

Setup of the motion for fitting the hour wheel and hand via reverse wheel and pinion.

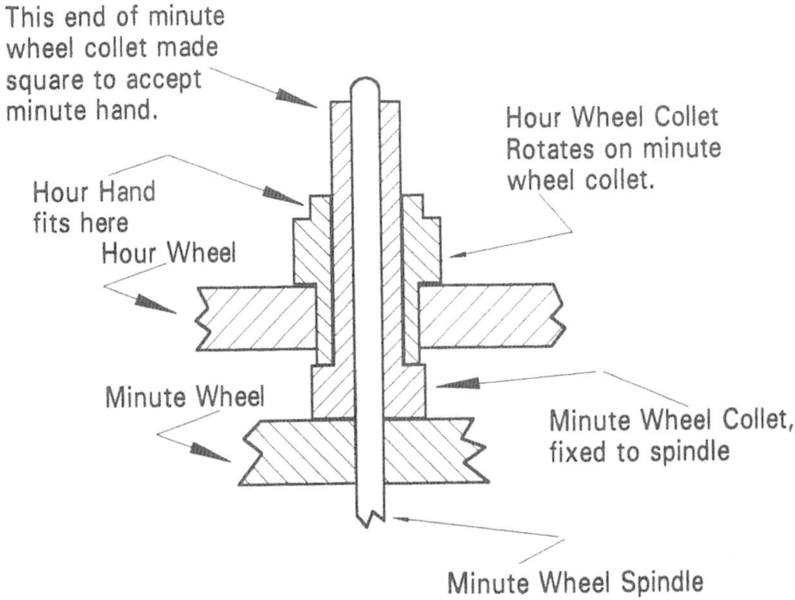

Sectioned view of the minute and hour wheel setup showing collet arrangement.

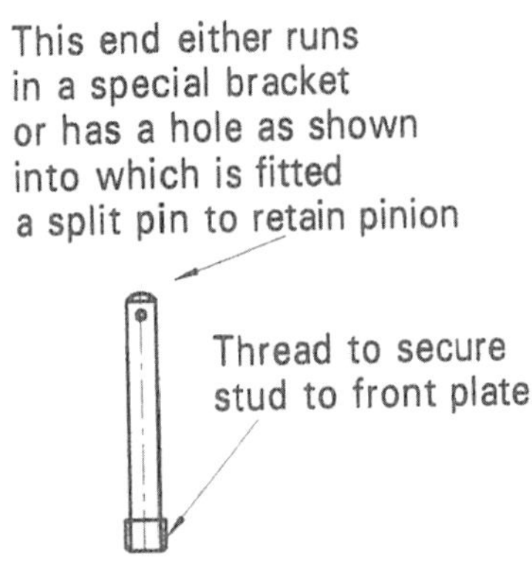

Detail of stud for reverse minute wheel.

The assembled motion work.

Chapter 7—Dividing

Accurate dividing is a prime requirement when cutting wheels, and there are a number of ways of doing this. Any reader who owns a dividing head will have no need of any advice on the subject and can go on to the section on cutting the teeth. Experience shows that, in general, dividing heads are not usually found in the home workshop; they tend to be one of the last pieces of equipment to be bought, and where they are found, in general they will be homemade. It is not within the scope of this book to go into how to make such a device, and in fact a full dividing head is not really necessary for clockmaking.

Many readers will be quite familiar with the use of lathe change wheels for dividing, and in some instances the idea will be quite good enough for our purposes. Consisting of simply a means of holding a lathe change wheel firmly in the mandrel of the lathe and a detent or pointer and that will fit exactly in the teeth of a change wheel and secured to a permanent point on the lathe, the idea has served model engineers well over the years. The main problem is the limited number of divisions that can be obtained. Few, if any, lathes will have change wheels with more than seventy teeth, which means that seventy is the highest number of divisions that can be obtained, and the divisions obtainable on the smaller wheels are extremely limited. In his book *Gears and Gear Cutting*, Ivan Law describes an excellent setup for compounding the dividing gears, thus giving a much wider range of divisions, as well as a mass of information on gear-cutting methods.

We must also consider the large number of lathe owners who do not have change wheels. For example, many of the smaller lathes that are quite popular for clockmaking do not have any; neither do those that are fitted with gearboxes. Most have a hollow mandrel that could be used to hold the wheel in position, but it will involve purchasing special gear wheels in order to use the system.

The alternative is to use dividing plates, which are more accurate than gears and can be easier to use. We tend to think of dividing plates as part of a dividing head, but this is not necessarily so; they can be secured directly to the lathe mandrel in the same way as a gear. In many ways, this is better than using a dividing head, where there is always the problem of backlash in the worm gear to worry about. Most dividing plates have a range of divisions on each plate, giving flexibility as well as accuracy. They can be purchased from model engineering suppliers or suppliers of materials for clockmaking. Generally speaking, it would be better to go to the latter, as the plates sold for general model engineering purposes are less likely to have the required number of divisions. Plates bought from clockmaking sources will often have exactly the right number of divisions for a complete clockmaking train, meaning that only one plate is required for the whole movement.

It is not all that difficult to make one's own division plates, provided that care is taken to ensure they are accurate. For best results, the plate must be of as large a diameter as possible and a series of circular lines should be

A wheel-cutting engine. Most horological devices such as this are referred to as "engines" rather than "machines." It is a simple device, consisting of a bracket holding a spindle, on which the cutter rotates via an electric motor that cannot be seen. The wheel is supported on the vertical spindle, which in turn is connected to a division plate. The use of a clock gauge ensures accuracy. Some readers may be struck by the similarity between this and the key-cutting machines that are commonly seen in marketplaces. On older such machines the cutters were operated by a handle, an idea that could well be adopted for making one's own wheel-cutting engine.

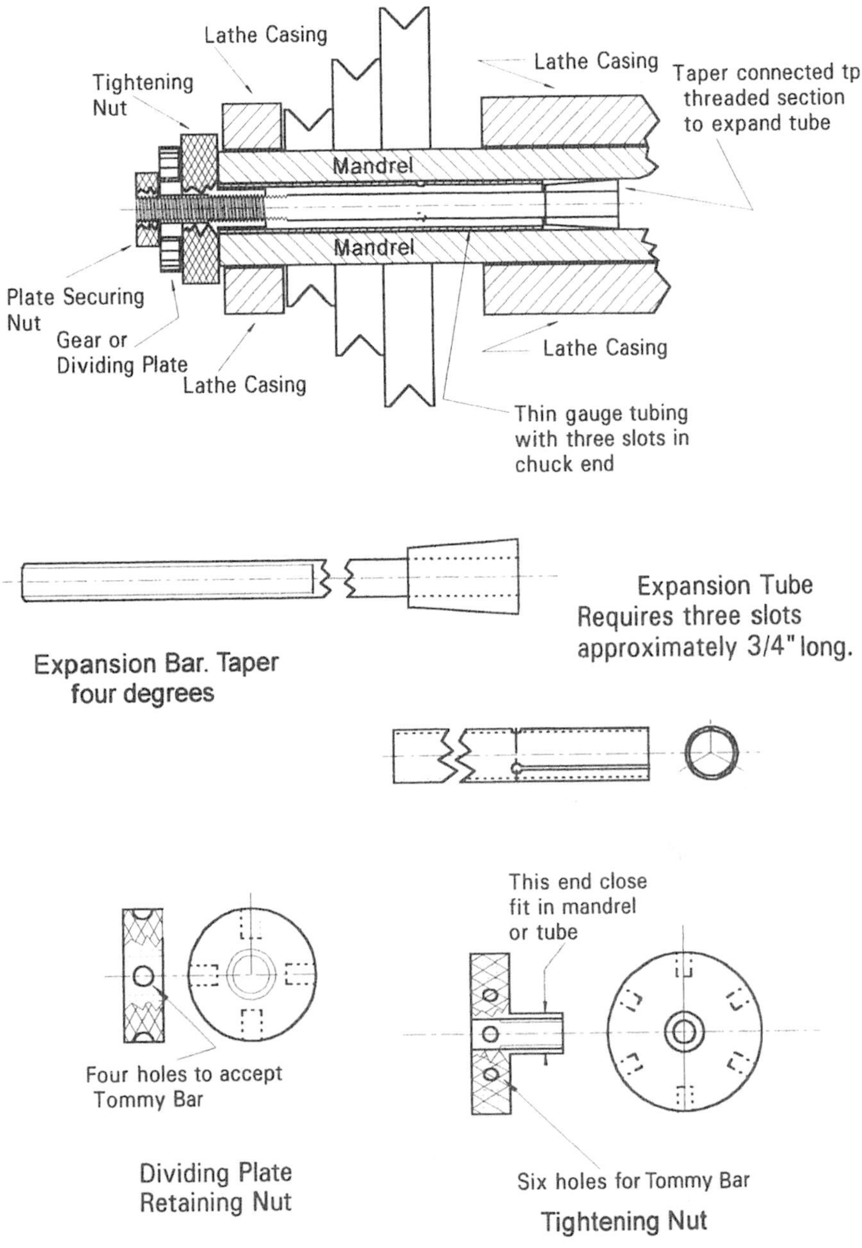

Adapter for holding dividing plates with lathe mandrel and parts for making unit.

scribed around it at known distances. It is of the utmost importance that these circles are accurate; if not, the divisions themselves will not be right. It is possible to scribe the circles accurately on the disk by using the cross slide graduations. Use a sharp pointed knife tool set at a suitable angle and allow it to just touch the disk while rotating the lathe by hand. A handle that can be secured in the mandrel is useful for this sort of work, as well as a number of other tasks that we come across when making clocks. Having scribed the circle, it is necessary to refer to a chord chart. It sounds obvious to just measure the circumference and then divide by the required number, but this does not work, because the distance measured between two points on the scribed circle, when using dividers, will be a straight line across the two points rather than around them. A chord table (see Appendix page 127) gives the figure for numbers of divisions, assuming the circle diameter to be one. To obtain the required figure, simply multiply the length of chord by the diameter of the scribed circle. Having established the length of the required division, take a pair of dividers that have nice, fine points on them and set the distance by reference to a micrometer or vernier gauge. Make a light center-punch mark

A simple method of dividing, provided that wheels with a suitable number of teeth are available, is to use the change wheels of the lathe and a detent.

somewhere on the scribed circle and mark off the divisions, starting and finishing at the center-punch mark. Lightly center punch each intersection. It is advisable to use a magnifying glass to set the dividers and to make the punch marks, in order to get the accuracy that is required.

Normal marking-out methods on a flat plane demand that all measurements are taken from a single datum, something that is not practical when dividing a circle. Therefore, the possibilities of multiple errors arise. For example, if the dividers are one hundredth of a millimeter oversize and there are a hundred divisions required, by the time the last one is reached a whole millimeter has been gained, and this is nowhere near accurate enough, so some means of checking therefore is required. Suppose we are to divide the plate into sixty divisions. Having set the dividers as suggested with a micrometer or vernier gauge and made the first indent on the line, mark off but don't spot three divisions. Preferably using another pair of dividers so that the original setting can be maintained, check the distance across the three marks. It should be exactly the chord figure for twenty divisions. Next mark off another three divisions and check again the whole distance, using the chord figure for ten. Any error in the original divider setting will now show

Homemade dividing plates such as these are practical propositions, provided that care is taken.

up and suitable adjustments can be made if need be. When it is right and not before, make the spot marks with the punch. Remember, the larger the plate and circle used, the greater the accuracy will always be.

Once the marks are made, it is simply a case of drilling through the plate, using a drill of the same size as the detent that it is proposed to use. It is essential to ensure the plate is flat and at exactly ninety degrees to the drill when it is being drilled, and in order to ensure this, it might be necessary to deburr each hole as it is made, so that the burrs do not tilt the plate when the next hole is made.

Make a series of such circles in the plate, preferably of the numbers that will be used in the clock train. Do be careful to ensure accuracy and measure at least twice, preferably three times, before making any mark, and then proceed as above. All the holes should not only be deburred on the side they break through, but also slightly countersunk on the face, to facilitate the detent entering when they are used. The detent should be a good fit in the holes, but not so tight that force is required. If it is tapered slightly, the good fit will be ensured, but the taper must be steep enough to prevent the detent from binding in the holes.

It is essential that the detent setup is secured to a position on the lathe in such a way that it will not move out of position. The actual detent that will locate with the holes in the dividing plate should be spring loaded if possible, largely as a means of saving time rather than for accuracy, which can still be achieved if it is a push-fit, as long as it can be locked in position. To keep unscrewing and then screwing up the device is very tedious, particularly if the division is for a large number of teeth, remembering that we are talking in many cases of wheels with a tooth count nearing a hundred.

Securing the Dividing Plate

The setup for securing the dividing plate is quite easy to make. A piece of thin-walled tube of a diameter suitable to be a good push-fit in the hollow mandrel is needed, and this should be about 4" or a hundred millimeters in length. If tube is not available, it will be necessary to drill or bore a length of solid bar. In one end, fit a threaded piece of bar, about ¾" or 20 mm in length. The diameter of the thread will depend on the diameter of the tube, but about three quarters of the bore is ideal.

A tapered plug is fitted to a length of threaded rod, or studding and a homemade nut, somewhat larger than normal so it pulls the tapered section up tight. The nut does not have to be hexagonal; it can be round and knurled, and for extra security a number of holes can be drilled around the periphery to accept a small tommy bar. The dividing plate or gear wheel is held in place with another similar but slightly smaller nut and with the addition of a special washer pulls the plate tight. It is essential that there is no play on the chuck-retaining setup that might cause some misalignment.

The device can also be used on a simple setup on the milling machine, which requires little more than a hefty angle plate with a hole to accept a mandrel to hold the lathe chuck. The other end is threaded so that a nut can be used to tighten the gear wheel or dividing plate that is in use and a detent can be fitted in a bracket screwed to the side of the unit. A screw is incorporated to lock the mandrel in position once the division has been set.

Dividing can also be carried out with a rotary table fitted with a means of holding a mandrel centrally. Usually this means making a device onto which a lathe chuck can be fitted. It is difficult to set up the system accurately, and it is better, therefore, to only use the rotary table for making division plates or, if no other system is available, with which to do the actual dividing.

A device for dividing when using the milling machine, it will accept either change wheels or dividing plates. It has the advantage that work can be carried out on the lathe and the chuck, complete with work transferred without any loss of accuracy as would occur if the work had to be removed and reset.

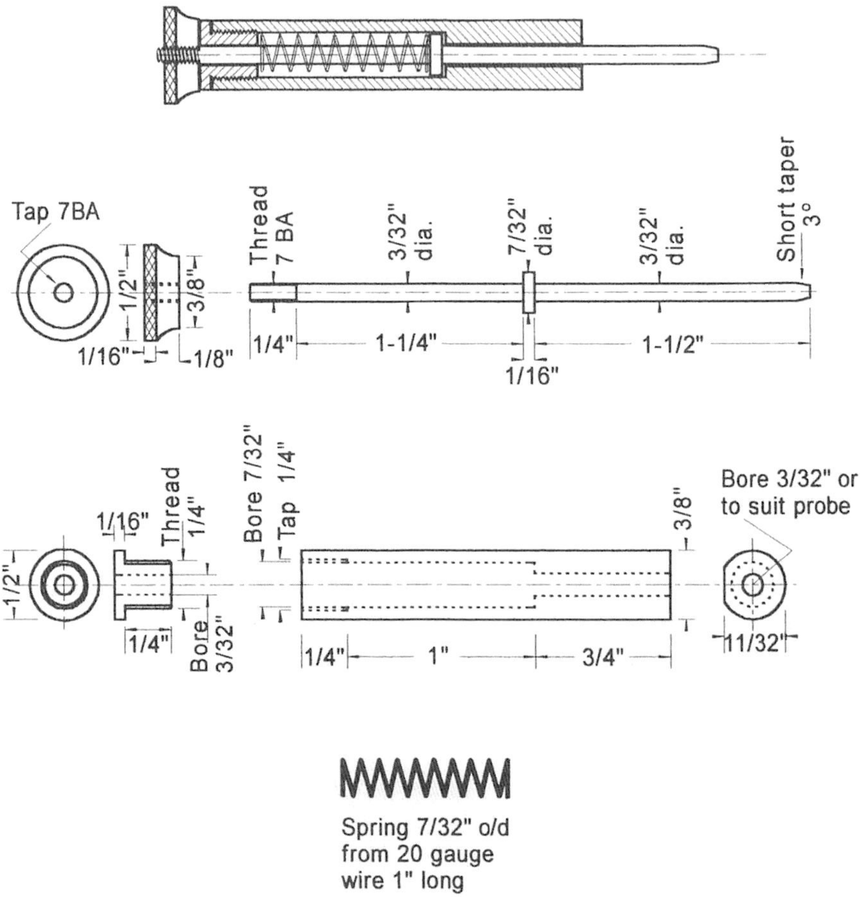

Detent for use with dividing plates. All mild steel, except the probe, which is drill rod. Sizes assume use of plates with 3/32" thickness.

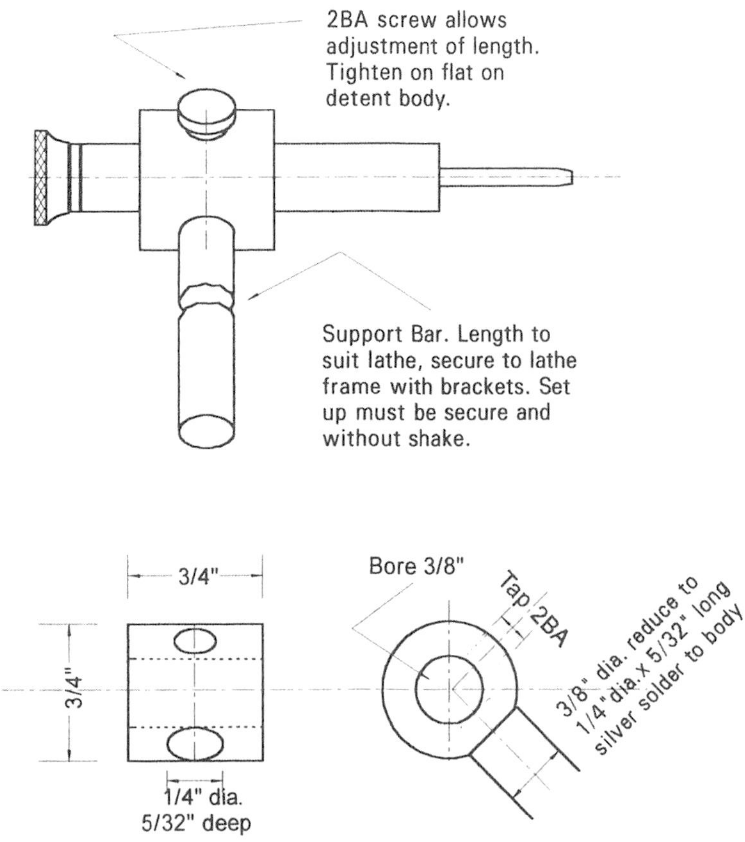

Setup for securing detent to lathe.

Chapter 8—Wheels and Pinions

In normal engineering practice, what is known as a "wheel" by the clockmaker would be referred to as a "gear," something with which we are all familiar, although the wheels in a clock differ in shape and form from the gears in an automobile. There are two types of teeth: involute and cycloidal, the former used in engineering, where it is more usual for the smaller gear to drive the larger one, and the involute form gives a better bearing surface for this purpose. In clockmaking, the cycloidal type is used because, in general, larger wheels are used to drive the smaller ones that are known as pinions, and the tooth pattern creates less friction. In all things to do with clockmaking, there is a constant battle to reduce both friction and weight of components in the search for efficiency. Engineering-type gears are described in Britain, America, and many other countries under the term *diametral pitch*, or DP for short. Measurement is found by dividing the number of teeth into the pitch circle diameter, which is a position where the teeth meet. The position is not visible when looking at a gear; only by taking various measurements of the gear teeth, etc., can the actual position be found. The module system uses the reverse of

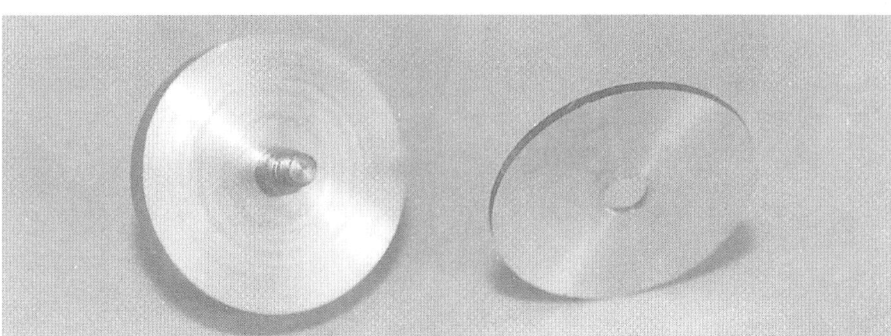

Large backing plates and supporting washers are essential when cutting the teeth on wheels. These are made of brass, which is quite suitable, but any material that happens to be on hand will do as long as it offers the required support.

this and is the number of teeth divided by the pitch diameter; it is entirely a metric system, whereas the DP system can be applied to either imperial or metric.

We are only going to deal with the module system because in clockmaking it is all that is necessary, although the clock repairer might well need to use both when repairing a very old clock. Most of the measurements that are needed are a multiplication of the module number. All necessary details for working out the wheel from the module as well as the module from the wheel sizes are included in the charts. The system is quite logical, and the larger the module number, the larger the teeth will be. Using a large number has the advantage that wheels are less easily interrupted in their operation by dirt, but large modules naturally are not suitable for small clocks such as carriage clocks.

A commercially made cutter. In this case it has fourteen teeth, but some have as many as sixty, leading to greater efficiency and a smoother cut.

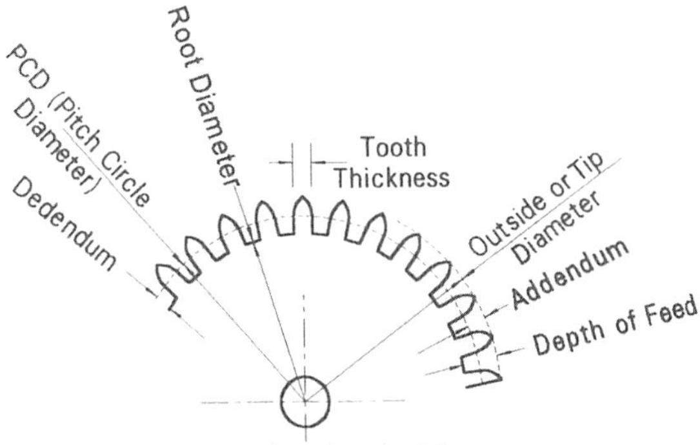

Terminology for obtaining measurements and module number of wheels

Wheel and Pinion Proportions

Wheels

- **Modules** = pitch diameter in mm divided by the number of teeth
- **Diametral Pitch** = number of teeth in a wheel per inch of diameter
- **Addendum** = distance from pcd to tip of teeth (1.35 × module)
- **Dedendum** = distance from pcd to base of teeth (1.57 × module 0.45 and 1.1 to 1.5 and 2 × module from 0.5 – 1) (Short form = 1.07 × module)
- **Pitch Circle Diameter** = number of teeth × module
- **Outside Diameter of Blank** = number of teeth + 2.76 × module
- **Root Diameter** = number of teeth minus 3.14 × module for modules 0.45 and 1.1 to 1.5; number of teeth minus 4 × module for modules 0.5 to 1.0
- **Tooth Thickness** = 1.57 × module
- **Addendum Radius** = 1.93 × module for modules 0.45 to 1.1 to 1.5 (includes short form)
- **Full Tooth Depth** = 2.95 × module for modules 0.45 and 1.1 to 1.5; 3.38 × module for module 0.5 to 1.0

Pinions

All dimensions are ratios of the module.

- **Pitch Circle Diameter** = number of leaves
- **Outside Tip Diameters** 6 = 7.71, 7 = 8.71, 8 = 9.71, 10 = 11.61, 12 = 13.61
- **Root Diameters** 6 = 2.5, 7 = 3.3, 8 = 4.2, 10 = 5.9, 12 = 4.8
- **Leaf Thickness** 6–8 = 1.05, 10–12 = 1.25
- **Addendum Radius** 6–8 = 1.05, 10–12 = 0.82
- **Tooth Pitch Ratio** 2–8 = ⅓, 10–12 = ⅕
- **Addendum** 6–8 = 0.855, 10–12 = 0.805
- **Dedendum** 6 = 1.75, 7 = 1.85, 8 = 1.9, 10 = 2.05, 12 = 2.10

Wheels

Except in exceptional circumstances, wheels are made of brass, and for most normal clockmaking purposes—and certainly when compared with normal engineering practice—the material used is of a very thin section. This creates its own problem when cutting teeth, since there is a tendency for the metal to bend away from the cutter if any attempt is made to advance it too rapidly, or if the material is not properly supported. Because of the thin material used, the wheel must always be supported on a collet when assembled on the arbor in order to give a greater support surface. If one reads any of the excellent books on clockmaking and repairing that were written many years ago, it will be seen that the usual method of holding the wheel on the collet was to rivet the end of the latter in such a way that the metal expanded and held the wheel firmly in position. Nowadays a far better idea is to use a retaining compound; not only does this give a secure bond, but with a small amount of heat the bond can be broken if required. The same applies when fitting the collet to the arbor: rather than use a force fit, a tiny drop of compound on a unit that is a good sliding fit will do the job far better.

The best brass to use is known as "compo" or "engraving brass" and is stocked by both clockmakers' suppliers and many model engineering suppliers as well. Suppliers of clockmaking materials often will be able to supply it as round blanks of the outside diameter required. Failing that, we are left with

Homemade single-point or fly cutters and a holder. One cutter is for wheels and the other for an escape wheel. They are quite efficient, as well as being cheap and comparatively easy to make.

two choices: it can either be cut from sheet or sliced from round bar. Although engraving brass is not available as bar stock, the metal will generally be found to be quite suitable. It is sheet material that is unlikely to be of the required quality. To cut from a sheet of suitable material, it will be necessary to first make a center and then mark off a circle a little larger than the outside diameter of the blank, which can be cut out, using a piercing saw or better still a mechanical scroll saw. The central hole is drilled, to the size of the collet on which the wheel will ultimately be mounted, ensuring it is at ninety degrees to the face and the disk and mounted on a suitable mandrel to be put on the lathe and machined to size and concentricity. It will be necessary to make disks that will fit on the outside of the blank to hold it firmly in position; because the diameter of the hole for the mandrel will be small, it is essential to avoid too much torque being applied by the tool when trying to machine the blank, and the washers will counteract this. If only the mandrel in the small hole is used, there will be two unwanted effects, the first and most obvious being that the blank will tend to bend as pressure is applied. The second, less obvious one is that no matter how careful one is when doing the machining, it will be

Cutting a wheel using the milling machine and a homemade dividing device. Note the support for the work to prevent distortion.

almost impossible to stop the blank from catching while the lathe continues to rotate, which will inevitably lead to some distortion of the hole, and no matter how slight this might be it will lead to loss of accuracy. If the work is supported and gripped by suitable washers, this should not occur.

Cutters

Gear cutters are available commercially and specialist clock suppliers should be able to supply a suitable cutter for any module that might be needed. These special cutters have the advantage of being small, which makes them particularly suitable for use on a small lathe. Although they are very expensive, they are made of high-speed steel and will last a lifetime. Cheaper cutters can be obtained; these are designed to fit on large horizontal milling machines, and although fitments can be made to enable them to be used on lathes and vertical milling machines, the setup is rather bulky. Frequently, too, these cutters are for involute gears rather than cycloidial, so check before buying.

Cutting a wheel using a Myford Lathe and a milling attachment from a small machine that has been adapted especially for the purpose. A fly cutter is being used, and in this instance there is not quite the same support as in the previous example. It is just about sufficient, but if the wheel had been the slightest bit bigger, a larger backing and washer would have been necessary.

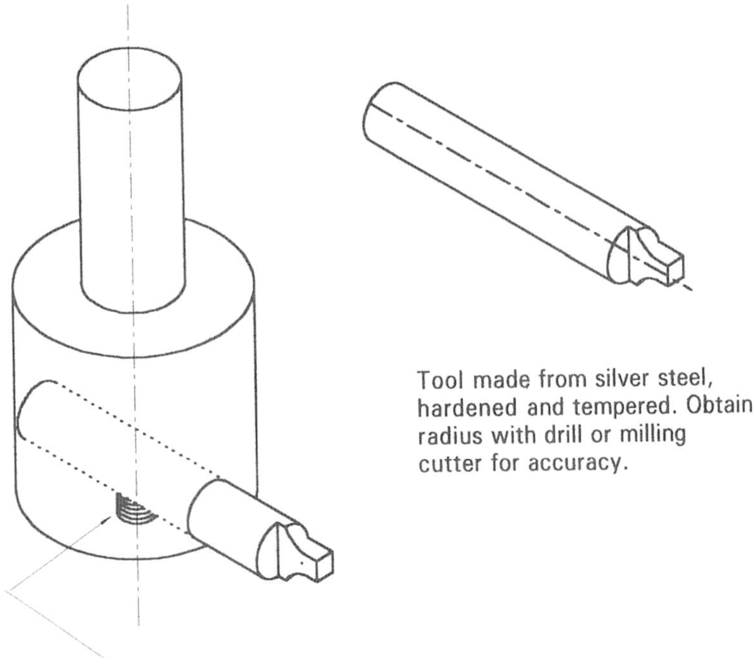

Method of making a single-point or fly cutter for machining the teeth on wheels and escapements.

Homemade Cutters

In Chapter 5, details for making cutters for escape wheels were given; it is also quite possible to make one's own cutters for wheels, using drill rod or gauge plate, which will be quite suitable for making the number of wheels required for a normal clock. When making cutters, remember that we are not going to make a tooth but to machine the gap between two teeth, shaping half of each in doing so. Commercial cutters are of the rotary type with the shape on the periphery, which is then divided into numerous cutting edges. Emulating this in the normal home workshop will be impossible, so it is best to aim at six or eight cutting edges, or a fly cutter with a single blade. To make a multi-tooth cutter calls for some type of form tool in order to obtain the accuracy needed; details of the required radius is as shown in the tables on page 83, 1.57 × the module. Therefore, before any calculations for teeth or cutters can be made, it is necessary to first find the module that is to be used. Generally this will be shown on the drawings, but there are instances where this may not be the case. For example, the author was interested in making a regulator, some details of which were shown in an old book. Only the outside diameters of the wheels and the numbers of teeth were given, and it was necessary to work out the module in order to take the project further.

The gap between the teeth of a wheel is the same as the thickness of a tooth and is one of the figures required; the second is the overall depth, and the third is the all-important radius. Two expressions crop up here, addendum and dedendum. They actually speak for themselves; addendum is a figure added to the point of the pitch diameter and dedendum is figure of the distance below it. The radius on a wheel is the shape of the addendum and is the only difficult part we are likely to come across when making a cutter, the other figures being quite straightforward. It is not going to be easy to get this radius with a file, and absolutely impossible with a grinding wheel unless one can be purchased that has been specially shaped. The best way of getting it accurate will be to drill a suitably sized hole and use that. The radii as shown in the chart do not work out to an exact size for which a drill is obtainable; for example, a 0.5 module has a radius of 0.965 mm, so to be exact we need a drill of 1.93 mm diameter. A standard size is 1.95 mm and that will be near enough for our purpose; if necessary, no great harm would come from using a 1.9 mm diameter drill, which is easier to obtain. For those who like to work in imperial measurements, 5⁄64" would be a suitable size.

Whether or not to make single or multipoint cutters is a matter of individual choice; many people are quite successful with fly cutters and see no need to go to the bother of making multi-point ones. With a single-point cutter, cutting the blank must be done much more slowly than with a multi-point one, although rotational speeds can be as high or even higher. There are numerous ideas on how to cut the blanks; generally it will be a matter of what equipment is available to the individual. The blank can be held in a setup on the vertical slide of the lathe, with the cutter rotating in the chuck, or the blank can be held and indexed in the lathe chuck or a collet and the cutter rotated on the lathe saddle. To do this, a device has to be made in which to rotate the cutter. Years ago, model engineers used to use a drive from an overhead belt; nowadays, with the ready availability of cheap, small electric motors, it is much easier to make the arrangement self-powered. An easy way is to make a bracket for a small model

A finished wheel and collet ready for polishing and assembly.

A commercially made pinion cutter.

Because of the small diameter of the pinion cutter, it has been necessary to taper the mandrel in order to give the shank sufficient strength.

Cutting a pinion, using the milling machine and a commercially made cutter.

maker's drill and to mount that on the vertical slide for height adjustment. There is little involved in doing so; any simple bracket will do as long as it will hold the drill firmly in place. Many small lathes are now available with milling attachments and these are ideal, as it means the blank can be held in the chuck and the cutter rotated on the milling attachment. Generally, these milling attachments are fixed to the lathe bed with a bracket, and an adapter will be required to attach it to the saddle so that it can be traversed, as it will be of no use in a fixed position. Owners of larger lathes might find it worthwhile to invest in one of these milling attachments for a small lathe and fitting it to the cross slide with a bracket. Those who have milling machines can use a simple indexing arrangement on the table, as described in Chapter 7, with the cutter mounted in the mandrel, and with all these alternatives a job that was once quite difficult has now become much easier.

It is essential that the cutter is set at the exact center height of the blank. Normal methods of obtaining center height are generally not accurate enough and it is best to use a scriber mounted in the chuck or collet of the lathe. In the case of the milling machine, a vernier height gauge can be used to set the cutter in relation to the blank. It is also essential to ensure that there is no shake or backlash on the mounting used for the blank, as this, too, will lead to loss of accuracy.

Pinions

The pinions, or small gears, of a clock set the clockmaker different problems to that of making wheels. Instead of easy-to-machine brass, drill rod is now the material to use, and rather than cutting through 1⁄16", or 1.5 mm thick material, the teeth, or "leaves" as the clockmaker calls them, will be at about ½", or 12 mm long. The shape of the leaves also differs slightly from the teeth of the wheel, with the result that the cutter used for wheels is not going to be suitable for the pinion. There are considerably fewer leaves or teeth than on a pinion, with numbers varying as a rule from six to twelve, and just occasionally for special purposes there are instances of four-leaf pinions, although there is no need to worry about those. The train, of course, dictates the number of leaves used; as we have already seen, that must work out right if the clock is going to work properly. Engineering practice tells us that the more leaves the pinion has, the smoother it will mesh and run with a gear, and there is no doubt that this is equally true when clockmaking. Therefore, ideally it would be nice if twelve-leaf pinions were used all the time, as it would make our clocks run smoother. Sadly, because this would mean very large wheels, this cannot be, and we are far more likely to be making six- or eight-leaf pinions, and in fact the more efficient twelve-leaf type is rarely seen at all. There are also three different profiles and, although in general only two are used, it is highly probable that a design might call for a different type. Pinions with a greater number of leaves use a different profile from those with a lesser number, the leaves being more rounded on the higher numbers.

In the past the clockmaker would make pinions from pinion wire, which was a long length of metal with the leaves already shaped. If there were

Where two or more identical pinions are required, machine a single length of metal and then part off sections to length.

too many leaves, they would simply cut one out and hammer the others to the correct position; the length that had been suitably doctored would then be cut off, polished, and taken into use. It was a highly-skilled process, and most model engineers will feel much more at home if they make their pinions by more conventional methods.

A similar system of measurement is used in the same way as for the wheels, but the leaves are much thinner than are the teeth of a wheel. This is to allow sufficient movement and clearance for the pinion to rotate with the wheel without creating too much friction, so while the width of a tooth on a wheel and the space in between is equal, in the case of a pinion the leaf takes one-third of the area and the space the remaining two-thirds. This applies to pinions with six to ten leaves; above that, the leaf occupies two-fifths and the space the other three-fifths.

Cutters for pinions can be purchased, but again, are very expensive, and homemade ones will be quite satisfactory for making a single clock. At first glance it would appear that making a suitable cutter is a very difficult proposition; while the teeth of a wheel are straight, those of the pinion taper inward. Because of the small number of teeth and the small diameter, this taper does not mean an increase in area at the root; instead there is a decrease, which actually makes the cutter easier to make rather than harder. The same methods are used to make the cutters as were used for the wheels. If a fly cutter is used, more care must be taken because of the length that has to be traveled. Therefore, a good supply of cutting fluid is essential to prevent the cutter, which must be fed very slowly, from overheating. The pinion must be supported at each end during cutting operations, otherwise it will flex away from the cutter, resulting at the very least in a bad profile, if not actually bending the work.

In order to get a good, smooth operation, it is essential that the leaves of pinions are given a good polish, otherwise they will be dragging on the teeth of the wheels. No matter how

The pinion after parting off.

careful one is when cutting, the end result will always result in a ragged finish of varying degrees. This may not be obvious to the naked eye, but look at it through a magnifying glass and it can be quite horrifying. Possibly the best method of polishing is to make a small profiled wheel from brass using the same cutter that was used to cut the pinion, coat this with a mild abrasive compound, and run it along the leaves until a suitable finish has been obtained.

While most people are quite capable of carrying out the work required to make a pinion, there are some who may feel the task somewhat daunting. In that case, it is worthwhile trying to make a lantern pinion, which, as the name suggests, looks like a lantern when finished. Except on replicas of old clocks, where the originals would quite possibly have had such a pinion anyway, the end result does not look as good as using the machine-cut version; however, it lacks little if anything in efficiency and so can be recommended for the beginner or for anyone who feels they do not yet have the ability to cut a normal pinion.

The pinion consists of two brass disks with a tube in the center that joins them, generally known as a "bobbin." The tube is designed to fit on the arbor of the wheel with which it is associated, and instead of teeth a series of rods connects the disks to each other. The drawing and photograph will explain the system far better than any words possibly can. Although the ends technically are disks, if making a lantern pinion, it is as well to use only one disk and to machine a length of solid brass, leaving one end

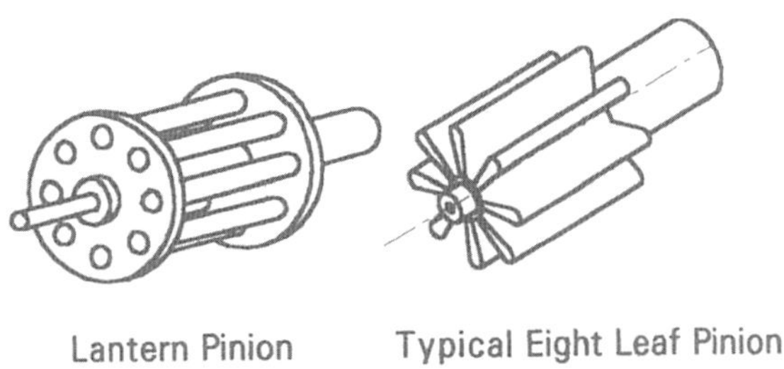

Pinions can be machined as part of the arbor if one wishes, rather than put on separately. The only disadvantage is that it is not then practical to use a depthing tool to mark out the hole positions.

to just over the diameter required and putting a step on the other. Do not part it off from the bar at this stage. Make a disk, again oversized, with a hole that will be a good fit for the step on the first piece. Solder the disk to the first piece; soft solder will do fine for this sort of work, but make sure the disk is square. Replace the bar in the chuck, and machine the outside diameter of the ends of the pinion to size, and at the same setting drill the central hole. This will need to be a good fit on the arbor, so ensure that the drill used is accurately ground, then just rub the cutting edges on a piece of emery cloth to take the very sharp edge off. Alternatively, drill the hole undersized and use a reamer to get it right.

Before parting off, index the holes and either spot them or, if facilities are available, drill them. If drilling, the holes can be passed right through to what will become the bottom plate. If not, part off and take the work to the drilling machine and drill through, making sure the work is perfectly square and that the drill also passes through square. It can be very difficult to see when a drill is wandering, particularly a small drill, which will bend. When all the holes are completed, pass through either lengths of hardened drill rod or special blued pivot steel that is available from clock material suppliers. Before doing this, put a spot of retaining compound on each. Finally, the ends of the pins will need to be ground off and the piece that has been used for chucking purposes also removed.

Of course, it is not possible to just pick any old size of pin of any circumference that takes one's fancy and finish with a pinion that is going to run with the rest of the train. The chart on page 83 shows how to find both the leaf thickness, which equals the diameter of the pin, and the pitch circle diameter, which is needed to place the pins. Therefore, if we want an eight-pin pinion for a number one module, the pitch diameter circle of the pins will

A lantern pinion under construction. The idea will work just as well as the open-top type and saves the cost of buying, or the effort of making, a cutter.

be 8 mm and the pin diameter 1.1 mm. Tooth depth also has to be considered, and the bobbin that holds the pins must have the small diameter, the same as that shown in the chart as root diameter, which in the stated instance is 4.2 mm. It can be slightly smaller for the sake of convenience if one wishes. Because of the length of the lantern—or any other pinion for that matter—there is no need to use a collet when mounting it on the arbor.

As we know, wheels and pinions are mounted on arbors, and the pivots are machined on the ends of these. Care has be taken to get a square edge when machining a pivot, as any taper means there could be a chance of it binding in the hole in the frame. Arbors can be made from drill rod, or special pivot steel in an attractive blue color that is already hardened can be bought. This is difficult to machine and the only way to use it without machining would be to keep it at its original diameter. To prevent it from moving through the holes in the plates, collars would have to be fitted. It will therefore be as well, unless one is very experienced, to use drill rod.

Chapter 9—Finishing

A well-finished clock movement is something that can be admired over and over again. Although, in general, we tend to think of polished brasswork, there is a great deal more to finishing than that alone. No matter how nicely polished the wheels and plates may be, the effect can be completely ruined by untidy work elsewhere. All polishing work should be carried out with a series of progressively finer polishing mediums, the type of which will depend on the original surface of the material being worked on. If the original surface is badly pitted, then work will have to start with various grades of sandpapers (abrasive papers) or cloths; their use should be kept to an absolute minimum because they can create more problems than they solve. Never use a piece of abrasive material that has previously been used on steel—on brass, minute particles of steel can become embedded and cause scratching of the surface.

Before work starts on polishing for appearance, it is necessary to carry out polishing to improve the working of the movement, and only then can the question of appearance be considered.

Throughout this book, the need to reduce friction and weight has been stressed, and first thoughts on polishing must be aimed at this. Machine-cutting methods invariably leave metal ragged and uneven, and no matter how careful one is, or how sharp the tools, to get the best from a clock, extra work is needed to remove these blemishes. This means trying to smooth the edges of the teeth on all wheels, including the escapement, and polishing other working surfaces. Obviously, when it comes to the teeth of wheels, a lot of care needs to be taken to ensure they do not lose their profile, so a piece of suitably shaped wood can be used in conjunction with a polishing medium, taking care to keep the wood at ninety degrees to the sides of the wheel. Escapement wheels in particular need attention, and it may again be necessary to make a suitably shaped piece of wood to get the best results.

Pivots

Pivots and the holes in which they are to run need attention, although the holes really should have been dealt with when they were made to ensure there was a good running fit with the pivot. Special finishing broaches are available for the purpose but anyone not wanting to invest in these can easily make a suitable tool from drill rod. Machine a short length to the same taper as the broach that was used to make the hole, and file the taper to half the diameter in the same way that one makes a D-bit. Remove any burrs from the edges, harden and temper to a dark straw color, and then just run the flat surface on a piece of very fine emery cloth or paper to give it a suitable edge. This will provide a nice polish to the holes, but do remember to keep it square when it is used.

The actual bearing surface of the pivot should be polished to as high a finish as possible. Special files can be bought for doing so, with an edge at an angle, which prevents destroying the square edge. As usual, there is no need to invest in such a tool, as our old friend,

To make the tool, start by making a suitable taper for the lathe. Set up the top slide for machining by laying a center between the tailstock and headstock and adjusting the top slide until a clock gauge reads zero throughout its length. Leave one end parallel.

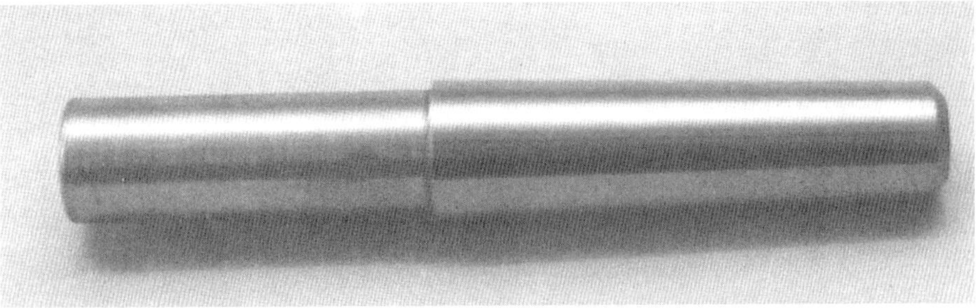

The finished Morse taper arbor.

Bore a hole to fit the parallel section of the taper in a block; any metal will do.

Drill and tap another hole below the first one and then a smaller one to accept a ball bearing farther along.

Drill and tap a hole across the block just above the first one.

Use either a slitting saw or hacksaw to make a slot into the first hole. If using a hacksaw, fit two blades in the saw to make a wide slot. If a slitting saw is used, either use one 3⁄32" wide or make two or three cuts to obtain the required width.

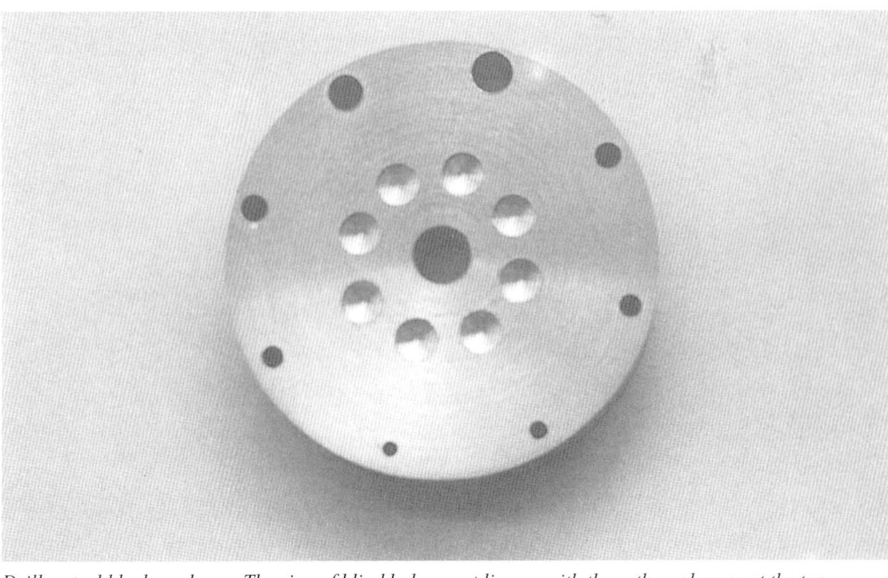

Drill a steel block as shown. The ring of blind holes must line up with those through ones at the top.

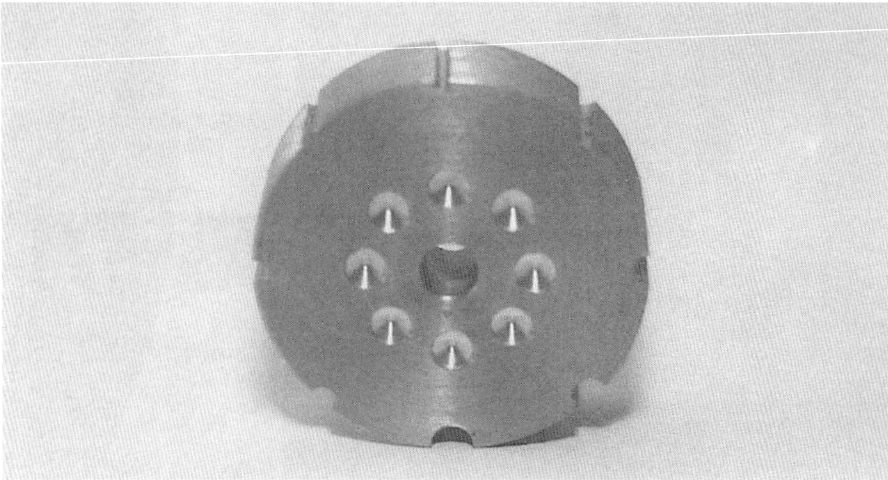

Put the block on a mandrel and machine the outer edge so that the holes are all equal to half the diameter.

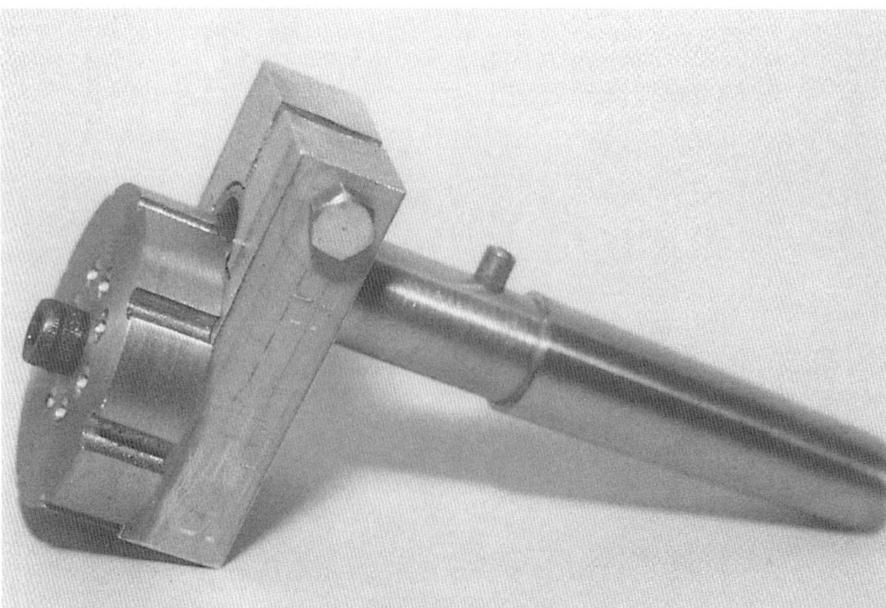

The completed Jacot tool ready for use. In this case, the mandrel is to be used for another purpose as well, hence the pin.

For polishing pivots, the Jacot tool is used. The varying depth of the grooves allows the pivot to rest at exactly center height, preventing the possibility of distortion.

a piece of hardwood, can be pressed into service. If the finish on the pivot is very bad, stick on some very fine emery paper as a start and use that, finishing as usual with a polishing medium. The biggest problem when polishing pivots is supporting them, and the best device for the purpose is a Jacot tool. (See photos on pages 98–103 and drawings on pages 106 and 107.) Consisting of a tailstock support of some sort, either a taper or bar that can be held in a chuck, a block is made that drops below center height, and fitted to that is a drum made of brass or plastic that has a series of grooves in the edge. These start life as holes, and then are machined to half their diameter so that when the block is rotated a groove can be set in a position to support a pivot. It is an easy tool to make and well worth the effort. However, there is an alternative, and that is to put a piece of wooden dowel in the tailstock chuck. From the headstock, drill a hole the diameter of the pivot to be polished, cut the dowel so that only half the hole diameter is left, and the pivot will rest in that while it is being polished.

Crossing Out

This is the horologist's term for reducing the weight of wheels by removing areas from the center, leaving a spoked effect, which can also look attractive. The shape of the spokes is a matter of personal choice, but they should be crossed out in such a way that nice square edges are left and when polishing is carried out those edges are maintained. Most of the work can be done with good quality needle files,

Finishing will include crossing out the wheels. The photograph shows one being marked out in preparation.

Needle files of various shapes can be used for the crossing out.

A better alternative can be a diamond-impregnated needle file, which leaves a more polished finish than the standard type.

The crossing out well under way: any design that suits the clockmaker can be used, and reading through this book, various ideas will be seen. Examples of other crossing out patterns can be seen on page 54.

stressing the need to buy quality files rather than cheap ones. It is far better to buy three or four good ones, rather than a wallet containing numerous cheap ones, the shapes of many of which would be unsuitable anyway. It is possible to buy escapement files, which are specially made for this sort of fine work and include special shapes. The author has made a small filing machine that is particularly useful for crossing out, as its use ensures that the edges remain square with the work supported at ninety degrees to the file.

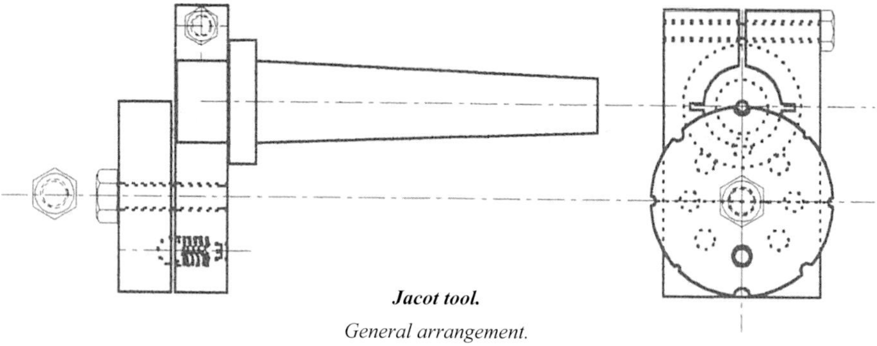

Jacot tool.

General arrangement.

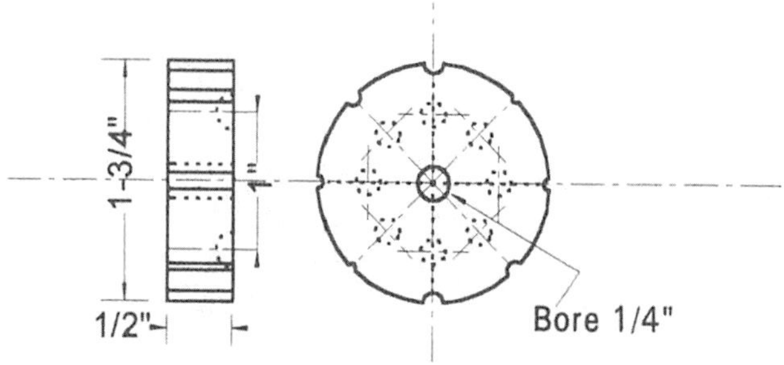

Front block

From mild steel or nylon, start at 2" diameter, face off, and drill for central pivot. Drill 8 holes at ⅞" centers; sizes to suit work likely to be carried out. Drill eight recesses 3⁄16" × 3⁄32" deep, round off by grinding using a 3⁄16" ball secured with adhesive to a rod. Machine piece to 1¾" diameter to break into holes at their centerline.

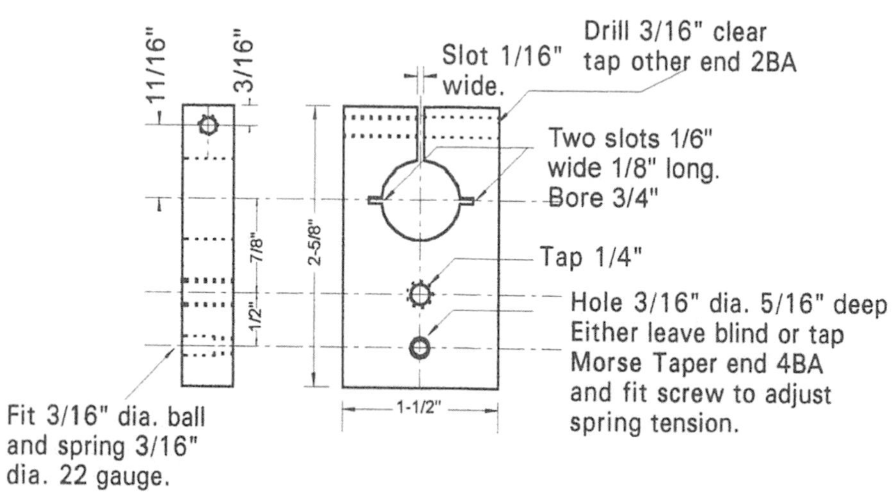

Rear block

From any suitable material.

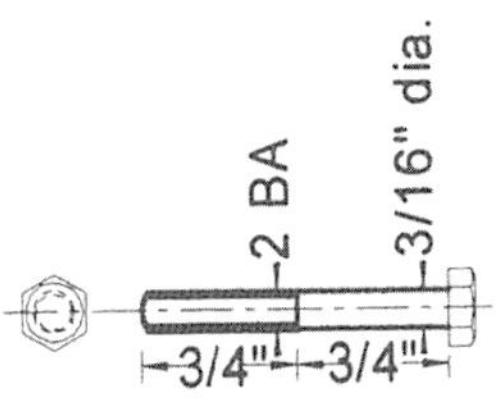

Clamping bolt

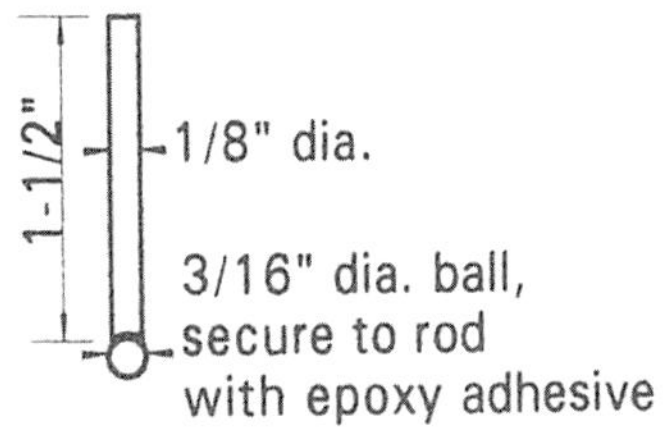

Recess grinding tool

Use with mild grinding paste such as toothpaste or scouring powder.

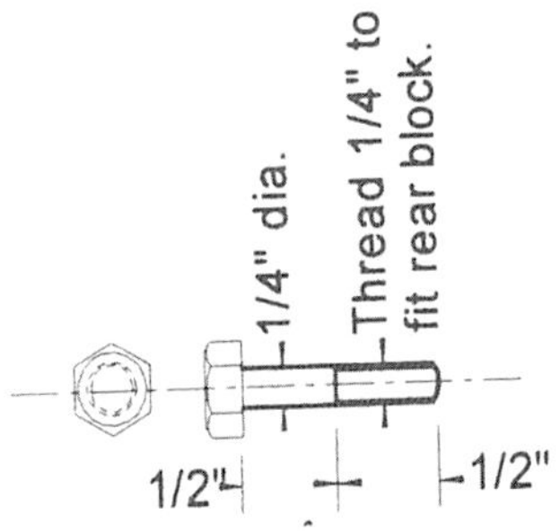

Front pivot screw

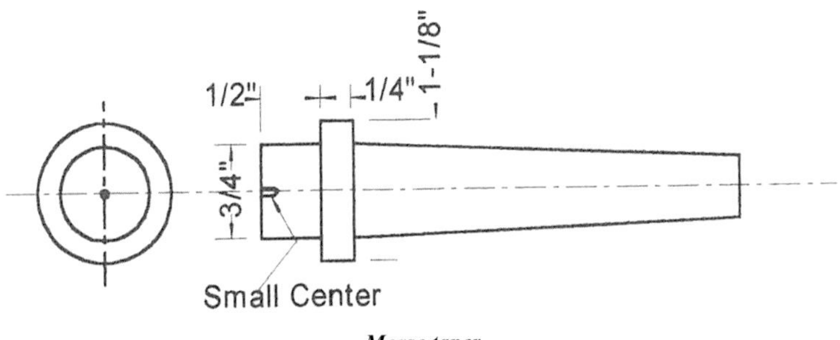

Morse taper

To suit lathe tailstock.

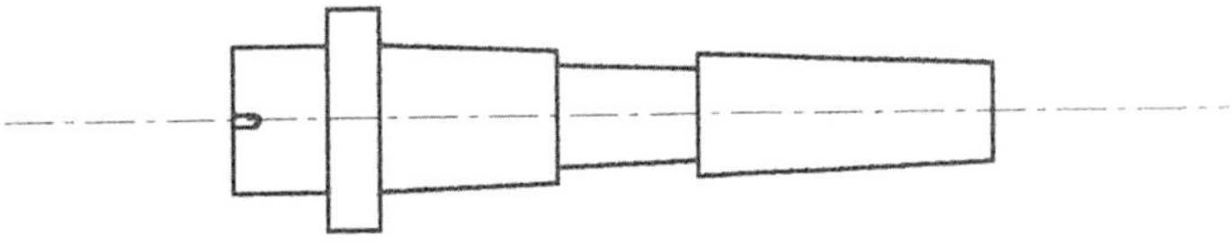

Alternative Morse taper

If in any doubt about the ability of a Morse taper to hold secure, machine a section from it, as above, to a depth of about 1⁄16" (1.6 mm).

Pallets

Obviously, it is nice to polish the visible surface of the pallets to make them look good, but most important of all is the need to impart a good finish to the working surfaces so they will mate smoothly with the teeth of the 'scape wheel. It is generally recommended that this be done with a wheel rotating in the lathe, while the pallets are supported on a hand tool rest. The polishing wheel is made of wood, and by supporting the pallet on the rest, the working surfaces can be contoured while remaining square to the sides.

Plates

First thoughts are that finishing plates is a comparatively simple task, but there are certain things that we need to look out for. All too often, a clock is spoiled by file marks along the plate edges, and care should be taken that all these are removed by draw filing, while at the same time ensuring that the edges are at ninety degrees and are kept square. Clamp the plates between lengths of angle to work on the edges, keeping the angle as close to the plate edges as possible, protecting the sides of the plates by putting paper between them and the angle before tightening up. Finish the edges with a very fine abrasive cloth wrapped tightly around a file, followed by a rub with a piece of square-edged hardwood with a liberal amount of a brass polisher spread on it.

Lay the plates on a flat surface for polishing the sides and use a block with a large surface area to do the work. Deep marks can be removed with sandpaper (abrasive paper); the type known as wet-and-dry is very good. Use the finest grade and wet it thoroughly, washing the residue off under a running tap. On fine scratches, use a piece of cardstock (card) stuck to the polishing block and soaked in a polishing medium. Ensure, when the plate is turned over, that the supporting surface is thoroughly clean and degreased before starting on the second side. It is very easy, when polishing the flat surfaces of the plates,

This small filing machine is ideal for crossing out.

to allow the abrasive material to cause a rounding off of the edges. This should be avoided at all costs. If space permits, use a large piece of cardstock (card) soaked in the polishing medium and laid on a flat surface, and work the plate on that, rather than the other way around. Use a figure-eight movement; the finished result should not be grained in any way but just have a highly polished surface. Polishing mops have their purposes, but generally their use results in rounded edges, which must be avoided.

Wrap the plates in clean cloth while awaiting reassembly.

Pillars

Pillars and the screws, if any, that hold them need particular attention. If the pillars are plain, there will be no problem, as they can be initially polished while rotating in the lathe. Finishing work should always be done along their length; no matter how carefully the work is done, witness marks invariably will remain on work done in the lathe. If the pillars have been shaped, we are faced with different problems, since machining marks are inevitable and these must be removed. Frequently tiny chatter marks are likely to be left in any recesses that have been made, and initially these will have to be removed with sandpaper (abrasive paper) while they are in the lathe. Although the use of a polishing mop on flat areas is not recommended, in the case of the pillars they can do a superb job, particularly as in most cases the last thing that will be wanted will be sharp edges.

Arbors

The materials to be used for making arbors has already been covered and for most people this means drill rod, which generally comes with an already finely ground surface. This finish can often be damaged in a variety of ways; for example, marks from chuck jaws are a common problem and are often caused by the work catching and remaining stationary while the lathe continues to revolve, resulting in either scoring or discoloration of the metal. It is easy to say, “Make sure it does not catch in that fashion,” but it is much harder to actually prevent it from so doing. If it does happen, the marks will have to be erased by polishing, and there are several schools of thought on how this should be done. The most popular method is to use emery cloth or a similar abrasive while the work is revolving; as with the pillars, it is very hard to disguise the polishing operation done in this way, and a fine finish can be obtained by working lengthwise. Once more, a range of polishing material should be used, becoming progressively finer as the finish improves. Here, too, is a job where a polishing mop can do a first-class job. The step that is machined to make the pivot should be treated with extra care. Firstly, the step must be at ninety degrees because, if it is at an angle, there is always a slight chance of it moving into the pivot holes and creating unwanted friction. It follows, therefore, that we do not really want to round the edges of the step-over when polishing, and one way to avoid this it to make a small ferrule to slip on the end so that emery paper, etc., will not tend to tip over the edge as it is moved along the length.

Materials

It cannot be stressed too much that coarse abrasive cloth or paper should never be used, and we should think only in terms of fine and extra fine materials. Where there are deep marks it is better to remove them with a Swiss precision file; these are available in a number of grades, and generally speaking Grade 0 will be needed for deep marks, moving to Grade 4 as the marks reduce in depth. Never use cheap files, because they will cause more problems than they will remove. In the long run, it will be cheaper to buy one good file that will outlast half a dozen cheap ones.

There are many proprietary materials available for imparting a very high finish, and they can be bought through good suppliers of horological materials; most will do a far superior job to anything that can be purchased elsewhere. No matter what materials are used, a perfect finish requires a lot of hard work, and unless it is carried out thoroughly, it will never be to a high standard. Don't be afraid to use a magnifying glass to examine the finish. Frequently looking at a part in an indifferent light will make it appear smooth and highly polished, but get it in a bright light, and particularly if that light is angled to the surface, and what once seemed to be perfect can look badly scored. If possible, look at it in daylight, which is far superior to any artificial light that is available.

Some people like to see blue arbors, and apart from using special blue pivot steel, there are other ways to achieve this. Most model engineering suppliers, all gunsmiths, and some suppliers of clock parts can supply the necessary chemicals to blue the steel. It is a simple process and the finished result can look very good, but in order to get the right result the steel must be highly polished in the first place. Chemical blueing will not remove blemishes, and must be regarded as a means of enhancing appearances rather than a quick fix to save a lot of polishing work.

Wheels

A normal engineering method of holding wheels for working on the flat surface would be to put pins in a piece of wood, adjacent to the edges, to prevent the work from moving with the action of the polishing. Such methods are rather too drastic for clockmaking purposes, since the edges of the wheels are likely to be at least marked very badly and at the worst irreparably damaged. Attempting to remove such marks would entirely change the shape of the wheel, so under no circumstances should this method be resorted to. It is far better to cut a small recess in a piece of wood, in which the wheel will fit without moving around, and with the edges standing out (proud) of the lip. This allows the polishing medium to be kept flat while the wheel is held firmly in position. Material can be saved by using the same piece of wood for all the wheels and making the recess gradually larger as the wheels increase in size.

A clock is judged on two things: its ability to keep time and its overall appearance, the latter only achieved by hard work and dedication, but in the long run is well worthwhile.

A finger plate is the best tool for holding wheels when working on them.

Chapter 10—Faces, Hands, and Cases

Having gone to a great deal of trouble to make a clock, it is beholden on the clockmaker to fit an attractive face to it, and there are numerous ideas that can be used for so doing. Before discussing these we should look at how the face will be fitted to the movement. In its basic form, after having been completed, the front face of the movement will consist of a plate, with four screws or bolts at the corners and a number of holes through which are sticking pieces of steel (the pivots) that go around when the clock is working. It is not, at this stage, the most attractive thing to look at. Anyway, it would not be possible to fit a face on it as it stands, so another plate called a false plate is put on the front. This has a large hole in the center that allows plenty of clearance for the hand collets and for the square on the end of the drum, which is to be used for

This unusual clock uses a chapter ring rather than a full face. Chapter rings are also common in other types of clocks.

A false plate is fitted in front of the motion and the face mounted on that.

winding. Four pillars keep the plate at a suitable distance and short extensions to these support a further plate; the screws into this are countersunk so they are not noticeable. It is known as the front plate and the face will be attached to it.

There are hundreds of commercially made faces in all sizes available for those who do not wish to attempt to make their own; they range from being very cheap to highly expensive. The expensive ones are works of art in their own right and are in most cases quite intricate in their design. They are made of metal and engraved or etched, not only with the numerals but also with various patterns. This is not to say that some of the cheaper commercially made dials are not attractive; a book like this cannot possibly offer a description of the whole wide range.

We are, anyway, concerned with making a clock, and for many, if not most, people this will include the face. We should start by separating faces into two parts: the full face and the chapter ring. The latter consists of a metal ring, of suitable diameter, printed with hours and probably subdivided into minutes. This is attached to a base, frequently of wood, and is quite attractive, particularly on larger clocks. The full face is exactly what the name suggests: a complete printed face as one unit.

Whichever type is chosen, the biggest problem will be the numerals. With care it is possible to engrave Roman numerals with the aid of a milling machine. It will be as well to first draw them full size on a piece of paper and then decide the best way to set about it. To form a figure I, it will be necessary to use a straight line with short cross pieces, and in order to get these accurate the graduations on the table must be worked to. Figures X or V will need to be a pair of lines, again with short cross pieces top and bottom; two sizes of cutter should be used on these

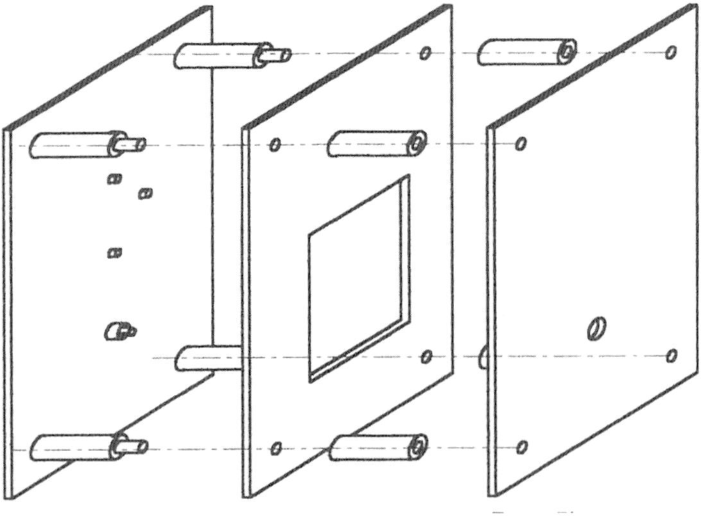

The fittings for false and face plate.

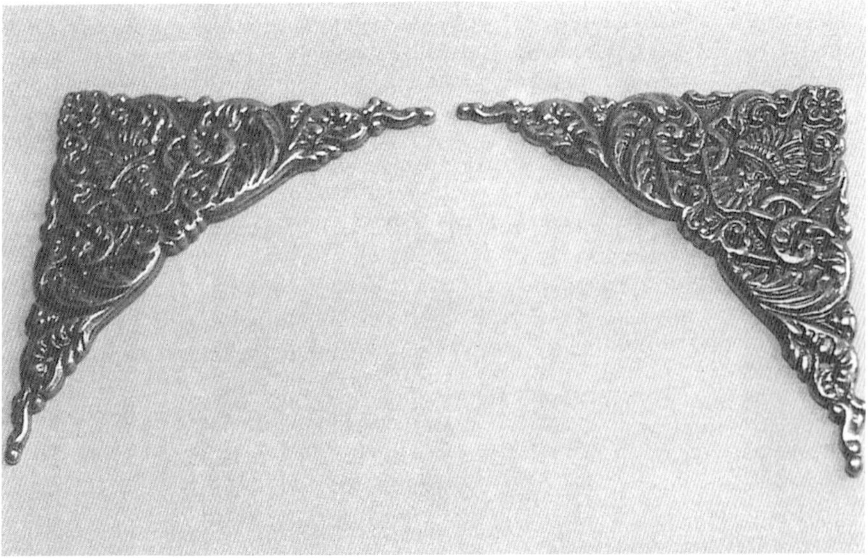

Decorative brass plaques for use with chapter rings, such as these and other designs, can be purchased and will enhance the appearance of any clock.

A small clock face, commercially made for use with a very tiny movement. Note the intricate patterns.

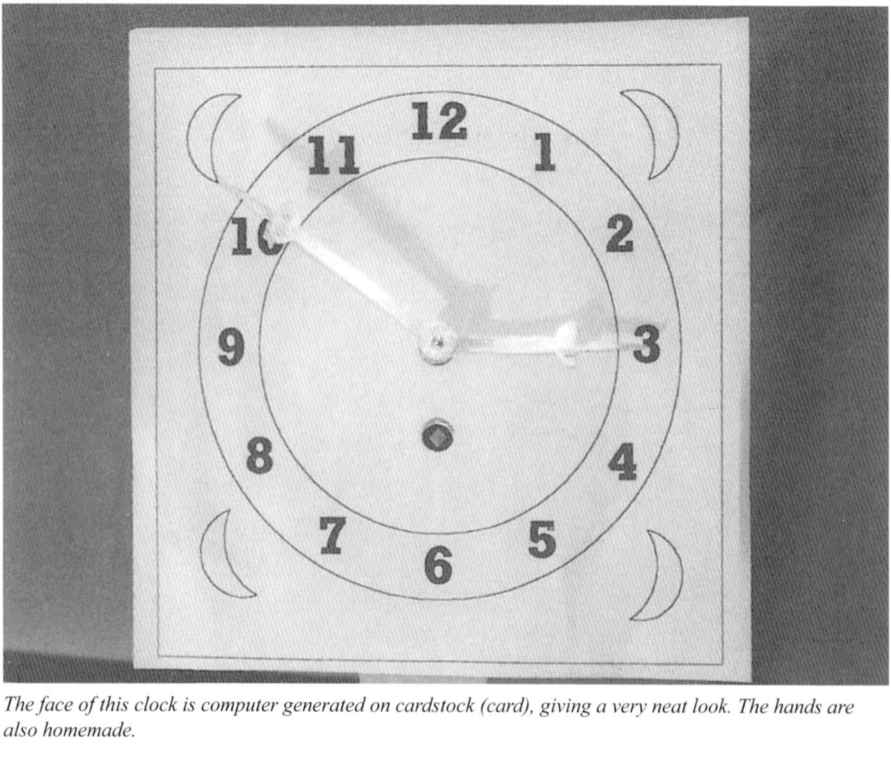

The face of this clock is computer generated on cardstock (card), giving a very neat look. The hands are also homemade.

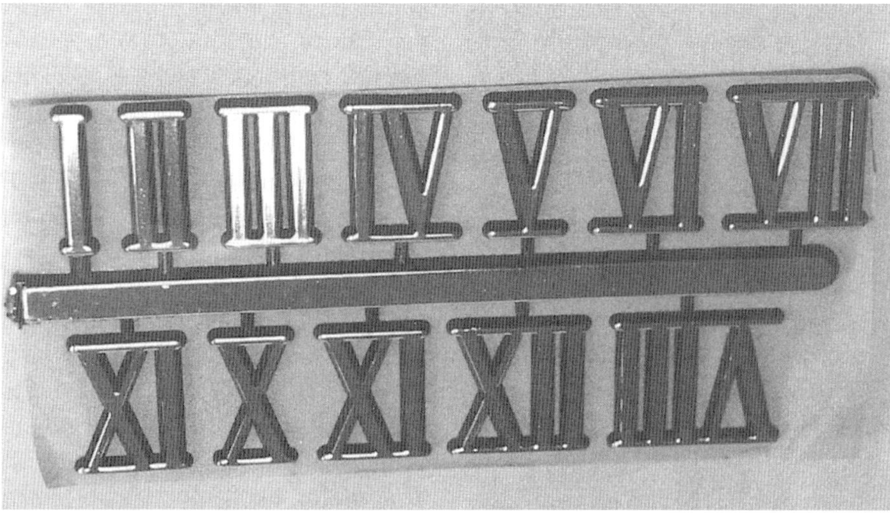

A set of numbers for making a clock face or chapter ring. They are self-adhesive and look very nice. Traditionally, four has always been shown as IIII on a clock dial. Various explanations have been put forward why this should be so, but they generally contradict each other. Sets of figures are available both ways.

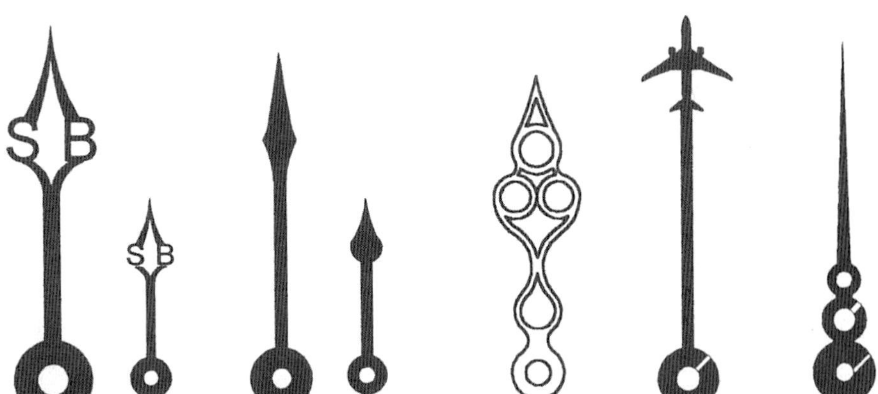

Hands can be made from thin sheet brass or steel. There are numerous standard shapes, but it is easy to create one's own designs.

Clock hands can be of any shape one wishes and are filed from brass like these.

figures to give an improved appearance. On small clock faces, normal milling cutters are likely to be too large, and dental burrs can be used instead. A point of interest is that a clock rarely shows the figure IV; four is traditionally shown as IIII. The finished numbers can be filled with black wax, which shows up well on brass or aluminum, whichever is used for the face or chapter ring.

It is possible to buy thin brass numbers that can be stuck onto any material and if one wishes could first be chemically blacked for appearance. It is also possible to buy self-adhesive plastic numerals with various finishes that can be stuck onto almost any surface. Transfers, which are specially designed for clock faces, are also readily available, and it is difficult to tell them from engraving. We are all aware of the wide range of rub-on transfers now available, and most art suppliers stock a considerably larger range than the average stationer. These do not look all that attractive when used for making clock faces, but if they are carefully applied to brass and then left to soak for twenty-four hours in an etching solution, as purchased at dealers in radio equipment, the brass around them will etch away, leaving the figures standing out (proud). Without removing the transfers, wash the face or ring in water and allow it to dry, but do not rub it to get it dry. Apply a chemical blacking solution to the metal, and when it has done its job, rub off the transfers. The result is a black face or ring with bright brass numbers. Be careful how the transfers are rubbed off, as the etching is not very deep, and if they are attacked with emery paper the numbers will be rubbed right away.

Readers can use their computers to produce their own designs for clock faces, which can be printed on thin cardstock (card) and stuck in place. Alternatively, they can be printed on a transfer sheet and this can be used directly on most materials. As far as design is concerned, a wide range of options is available if a computer is used. Anyone who has completed a

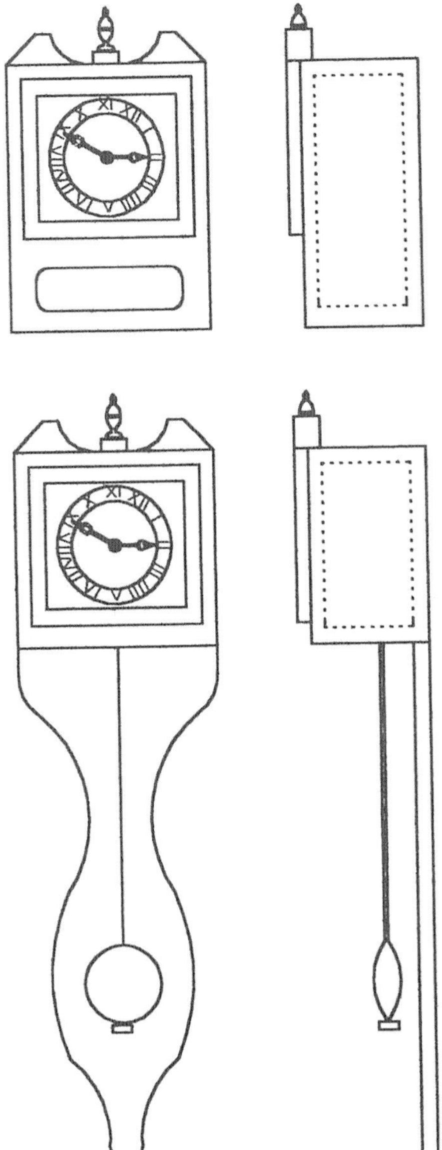

The same principle as described for a long case clock can be used for making cases for a bracket clock (above) or a wall clock.

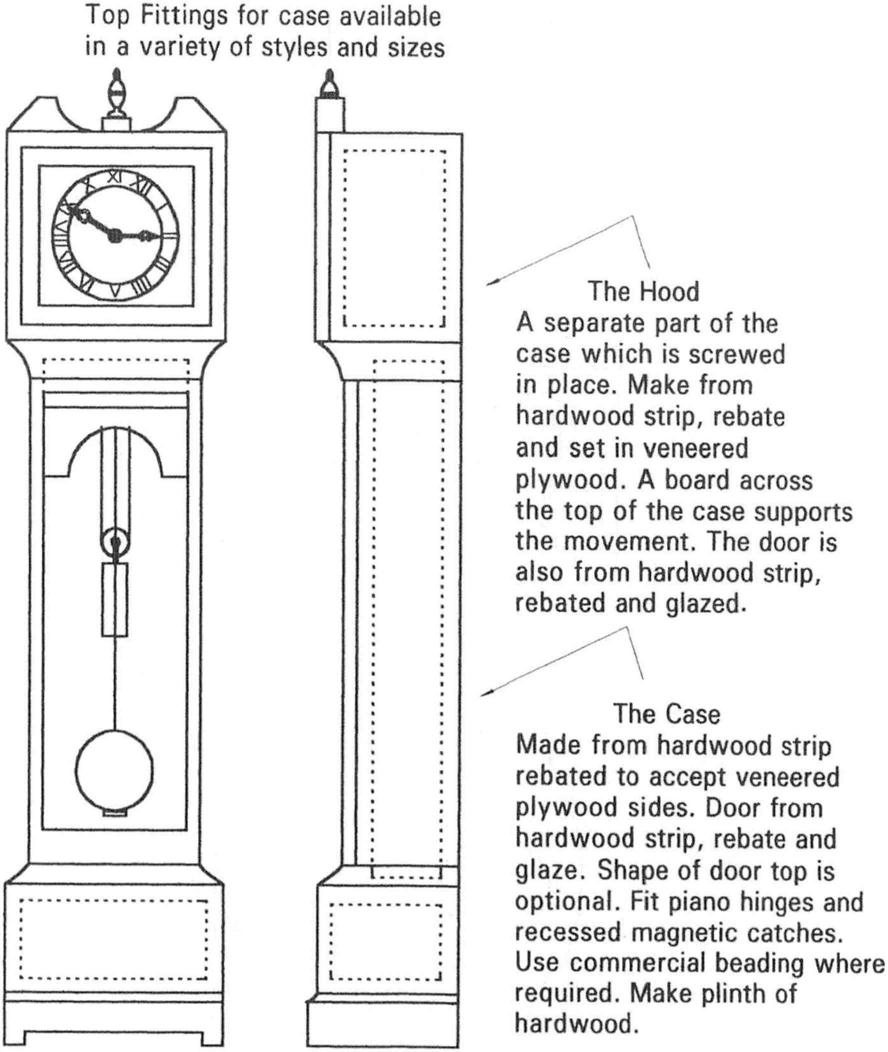

Makeup of a typical fabricated case for a long case (grandfather) clock.

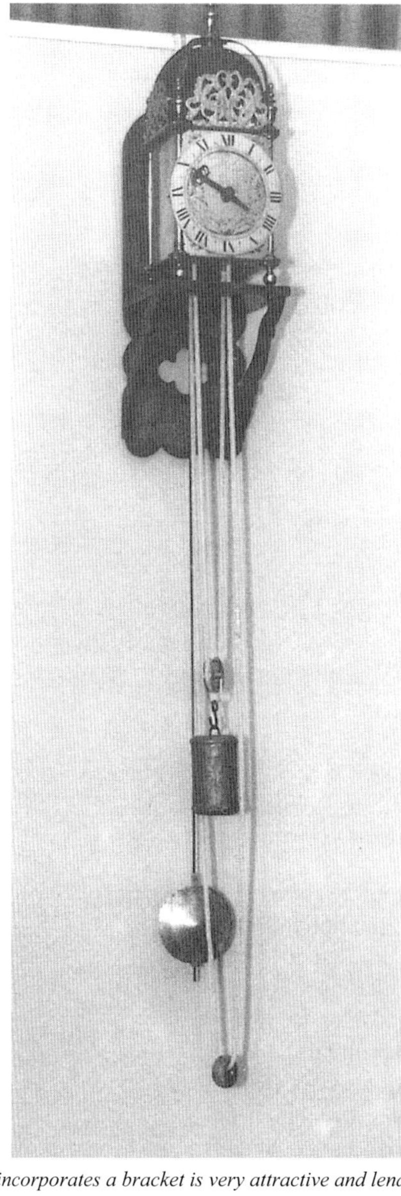

A wall clock where the case incorporates a bracket is very attractive and lends itself to embellishments such as the brasswork in this instance.

Most clock cases are made of wood. This example of a carriage clock, however, is made of brass plates soldered together. When polished it is very attractive.

movement would be advised to take their time when considering the face, and, in fact, it is a good idea to draw one on paper, with or without the aid of a computer, and set it temporarily in place. A day or two later, draw another one and see how that looks. Keep experimenting until a good idea of how the finished article will look is obtained.

Hands

Hands can be cut from thin brass or steel; once again, the less weight they have, the better. There is a wide range of more or less standard designs, and in addition, there is no reason why personal ideas cannot be used. For example, they could reflect a hobby or other interest, or perhaps something to do with the family. All these things give individuality to a clock and make it personal. It is as well to draw them first of all on paper, which is stuck onto the material from which they are to be made. They are then fretted out using a piercing saw and needle files. The fastener will depend on the design of a particular clock, and it may or may not call for a small piece of tubing to be soldered to the hand. The finished item can be chemically blacked or blued if one wishes; alternatively, it could be painted. Once again, it must be said that there is a very wide range of both designs and sizes available commercially, although the method of fastening may have to be altered on bought ones.

Cases

The making of a case generally calls for a different set of skills since, in the main, the work will be with wood. There are exceptions. Skeleton clocks are usually displayed in glass domes, and it is doubtful whether many people would wish to tackle such a task. Sometimes a clock will be displayed in a glass case, the making of which should not be beyond most people. There are a number of ways to go about the task; one of the easiest is to order the glass from the local glazier, specifying that the edges must be polished. This can be held together using a clear silicone and nothing else, and looks attractive since there is no wooden beading to obscure the view of the clock. The base

Faces like this are easily generated on a computer. They can either he printed on cardstock (card) and stuck in place, or produced as transfers. Embellishments can be added if wanted. Note the difference in position of the numerals. Roman point toward the center while Arabic are always upright.

on which the case stands should be of hardwood and have a groove cut in it for the glass case to slip into.

The more traditional glass case consists of thin wooden beading with the glass let into it. Again, hardwood should be used, and the groove for the glass can be made with a milling cutter if a router is not available. The corner joints can be mitered to improve appearance; once again, it is worth having the edges of the glass polished before making the case.

Generally speaking, we think in more traditional terms when considering cases, and the choice of wood becomes important. Ideally the case should be made of hardwood planks, but these are getting more and more difficult to obtain. For long case clocks, some sections could be made of veneered blockboard, but the problem of obtaining suitable materials is becoming very difficult indeed. One answer is to use hardwood strips and fit a good quality veneered plywood in machined grooves; it is hard to tell the finished result from solid wood. It should be possible to get suitable materials to make smaller clock cases from a lumberyard (timber merchant) that deals in hardwoods, although it may mean dealing with a company some distance from where one lives because there are not many of them left these days. Information on where to obtain wood can be obtained from the advertisements in magazines dealing with woodworking or clocks.

Cases can be finished with one of the modern varnishes, although traditionally they have always been French polished. Once again, the best advice is to contact advertisers in specialty magazines for help and information before deciding the best way to go about it. It is best to avoid DIY stores when thinking of materials; their stock is aimed at a mass market and for a different purpose, and is unlikely to be suitable for this sort of work.

Many clock cases are finished with fancy, shaped beading and ornaments which can be obtained from suppliers of clockmaking equipment; although mostly wood, some of these embellishments are brass, and either way they do add the finishing touches to a clock case.

Appendix

Formulas

Going train formula for checking correct wheel arrangement and for finding length of power chord required.

$$\frac{\textbf{No. of teeth in center wheel} \times \textbf{No. of teeth in 3}^{\textbf{rd}}\textbf{ wheel}}{\textbf{No. of teeth in 3}^{\textbf{rd}}\textbf{ pinion} \times \textbf{No. of teeth in escape wheel pinion}} \textbf{ or } \frac{64 \times 60}{8 \times 8} = 60$$

As the center wheel pinion has eight teeth and the great wheel ninety-six, the wheel rotates: $\frac{96}{8} = \textbf{12 hours}$

And if the diameter of the barrel is 2", the cord will be unwound.

$\pi \times 2 = 6.28$" in the same period. As the cord is double, the weight falls through only half the distance that it unwinds from the barrel; ie, 3.14" in twelve hours or just over 6¼" every full day of twenty-four hours. That is the equivalent of 4'2" in eight days, which is about the maximum that most people are likely to want it to unwind. It amounts to about sixteen turns around the wheel, and so there should be about seventeen coils on the drum.

Pendulums

The mathematical calculation for timing a complete oscillation of a simple pendulum is:

$$\textbf{Time} = \pi \sqrt{\frac{\textbf{length}}{\textbf{gravity}}} \quad \textbf{in feet}$$

$$\textbf{or } t = \sqrt{\frac{l}{g}} \quad \textbf{where } \pi = 3.14159, \textbf{ gravity} = 32.19$$

To calculate the length of a pendulum required for a given train of wheels, the total number of teeth in the center, third, and escape wheels are multiplied together and then multiplied by two. They are then divided by the number of leaves in the pinions of third and escape, multiplied together.

For example, Center Wheel = 64t - Third Wheel = 60t - Escape Wheel = 30t Pinions are both 8 leaf.

$$\frac{64 \times 60 \times 30 \times 2}{8 \times 8} = \textbf{3600 beats per hour}$$

$$\frac{3600}{60} = \textbf{60 beats per minute}$$

Chord Tables

To divide a circle into even sections, use the table below. The figures given are for a diameter of one. To find required figure, multiply length of chord for the number of spaces wanted by diameter of circle to be divided.

Number of Spaces	Length of Chord	Number of Spaces	Length of Chord	Number of Spaces	Length of Chord
3	0.8860	36	0.0872	69	0.0455
4	0.7071	37	0.0848	70	0.0449
5	0.5878	38	0.0826	71	0.0442
6	0.5000	39	0.0805	72	0.0436
7	0.4339	40	0.0785	73	0.0430
8	0.3827	41	0.0765	74	0.0424
9	0.3420	42	0.0747	75	0.0419
10	0.3090	43	0.0730	76	0.0413
11	0.2817	44	0.0713	77	0.0408
12	0.2588	45	0.0698	78	0.0403
13	0.2393	46	0.0682	79	0.0398
14	0.2225	47	0.0668	80	0.0393
15	0.2079	48	0.0654	81	0.0388
16	0.1951	49	0.0641	82	0.0383
17	0.1838	50	0.0628	83	0.0378
18	0.1736	51	0.0616	84	0.0374
19	0.1646	52	0.0604	85	0.0370
20	0.1564	53	0.0592	86	0.0365
20	0.1590	54	0.0581	87	0.0361
22	0.1423	55	0.0571	88	0.0357
23	0.1362	56	0.0561	89	0.0353
24	0.1305	57	0.0551	90	0.0349
25	0.1253	58	0.0541	91	0.0345
26	0.1205	59	0.0532	92	0.0341
27	0.1161	60	0.0523	93	0.0338
28	0.1120	61	0.0515	94	0.0334
29	0.1081	62	0.0507	95	0.0331
30	0.1081	62	0.0507	96	0.0331
31	0.1012	64	0.0491	97	0.0324
32	0.0980	65	0.0483	98	0.0321
33	0.0951	66	0.0476	99	0.0317
34	0.0923	67	0.0469	100	0.0314
35	0.0896	68	0.0462		

Common Clock Trains

Center Wheel	3rd Pinion	3rd Wheel	'scape Pinion	'scape Wheel	Vibrations per Minute	Length of Pendulum
112	14	105	14	60	60	39.14"
96	12	90	12	30	60	39.14"
80	10	75	10	30	60	39.14"
64	8	60	8	30	60	39.14"
75	8	60	8	32	75	25.53"
80	8	72	8	30	90	17.39"
108	12	100	10	32	96	15.28"